Pragmatic Circuits:
Frequency Domain

Pragmatic Circuits: Frequency Domain

William J. Eccles

ISBN (13 digits): 978-3-031-79748-4   paperback
ISBN (13 digits): 978-3-031-79749-1   ebook

DOI: 10.1007/978-3-031-79749-1

A Publication in the Springer series
SYNTHESIS LECTURES ON DIGITAL CIRCUITS AND SYSTEMS #3

Series Editor: Mitchell Thornton, Southern Methodist University

Series ISSN: 1932-3166   print
Series ISSN: 1932-3174   electronic

10 9 8 7 6 5 4 3 2 1

Pragmatic Circuits: Frequency Domain

William J. Eccles
Rose-Hulman Institute of Technology
Terre Haute, Indiana, USA

SYNTHESIS LECTURES ON DIGITAL CIRCUITS AND SYSTEMS #3

ABSTRACT

Pragmatic Circuits: Frequency Domain goes through the *Laplace transform* to get from the time domain to topics that include the *s-plane*, *Bode diagrams*, and the *sinusoidal steady state*. This second of three volumes ends with a-c power, which, although it is just a special case of the sinusoidal steady state, is an important topic with unique techniques and terminology. ***Pragmatic Circuits: Frequency Domain*** is focused on the frequency domain. In other words, time will no longer be the independent variable in our analysis. The two other volumes in the ***Pragmatic Circuits*** series include titles on ***DC and Time Domain*** and ***Signals and Filters***.

These short lecture books will be of use to students at any level of electrical engineering and for practicing engineers, or scientists, in any field looking for a practical and applied introduction to circuits and signals. The author's "pragmatic" and applied style gives a unique and helpful "non-idealistic, practical, opinionated" introduction to circuits.

KEYWORDS:

frequency domain, s-domain, a-c power, sinusoidal steady state

Contents

CHAPTER 5

Laplace Transforms: *s* is a Big Help

Laplace transforms. You've probably learned their ins and the outs in some math class some-where. In the math class, you learned a lot more about them than you'll need here, so we aren't going to stretch your knowledge of them at all. The surprise, if there is one, may be that Laplace transforms are actually useful in circuit analysis and design!

How useful? Well, we base the study of circuits in the *frequency domain* on Laplace trans-forms. They are important enough that we even find ways of circumventing some of the tedium of using them. The next three chapters will deal in some way with the transform and its appli-cations to circuits.

While I think of it, this is a good place to point out a very basic notion about our use of the Laplace transform. We talk about working in the "frequency domain," but we sometimes don't make clear where this comes from. Here's the integral that gives us the Laplace transform of $f(t)$:

$$F(s) = \int_{0^-}^{\infty} f(t)e^{-st}dt.$$

We start with $f(t)$, which is, simply, a function of time. Remember from math that t, the inde-pendent variable, says we are operating in the *time domain*.

What about the independent variable in $F(s)$? When we do the integration for the Laplace transform of $f(t)$, we "integrate out" the variable t. The result is a function of s. Now s is the independent variable and we are operating in the s *domain*.

So far so good. Now let's look at s itself. How about its units? Consider the exponent of the exponential e in the integral. That exponent is $-st$. Exponents of e must be unitless, so if t is to have the common unit of *seconds*, then s must have the unit of "/second" or "per second." That's *frequency*! (Yes, I know, *what* per second? The bland, unitless quantity *radians* will have to do.) So s is the frequency and $F(s)$ is a function in the *frequency domain*.

So as you talk about circuits and listen to others, you'll find that, part of the time, you want to know about the circuit and its time response. Then you'll be working in the *time domain.* At other times you'll care only about the circuit's frequency response, so you'll be working in the *frequency domain.*

Have you ever looked at the response curve of a loudspeaker? Or the characteristics of a stereo? Then you've seen graphs in the frequency domain. We'll learn about these graphs ("Bode diagrams") in Chapter 7. The data will show such things as, "This speaker's response is flat from 300 Hz to 10 kHz." That's *frequency-domain* data.

In this chapter I'm going to start with some work in the time domain, which means differential equations. Then I am going to transform these to the frequency domain via the Laplace transform and see what happens. After that, we'll learn that we can convert the circuit itself to the frequency domain without going through differential equations and the Laplace transform.

5.1   TIME DOMAIN EXAMPLE

Solutions of circuits problems in the time domain generally require differential equations, at least if the circuits contain energy-storage elements. And our circuits generally do—after all, how much fun can you have with resistors after the 67th problem?

In this section I am going to solve one circuit by writing and solving the differential equations. I'll write the equations in two ways, however. One will be the formal nodal analysis method that we already know. But when there are energy-storage elements, another method often works better: state equations. The result will be a set of *first-order* differential equations. Nodal analysis generally yields equations of higher order derivatives or equations containing integrals.

5.1.1   Solution via Nodal Analysis

Let's start with the circuit of Fig. 5.1, which has two energy-storage elements in it.

We are told that both energy-storage elements have initial energy, expressed as

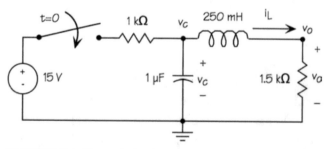

FIGURE 5.1: Example in time domain.

$$i_L(0) = 10 \text{ mA},$$
$$v_C(0) = 5 \text{ V}.$$

We are to find $v_o(t)$ for $t \geq 0$.

Let's use nodal analysis, writing KCL at the nodes labeled v_c and v_o.

$$\frac{v_C - 15u(t)}{1000} + 10^{-6}\frac{dv_C}{dt} +$$

$$\frac{1}{0.25}\int_0^t (v_C - v_o)dx + 10 \times 10^{-3} = 0,$$

$$\frac{1}{0.25}\int_0^t (v_o - v_C)dx - 10 \times 10^{-3} + \frac{v_o}{1500} = 0.$$

While Maple can solve this set of equations (which must include the initial value of v_C), it will do so only if the "laplace" option is added to the **dsolve** statement. If we must solve by hand, we generally need to differentiate the equations to eliminate the integrals:

$$\frac{1}{1000}\frac{dv_C}{dt} - \frac{15}{1000}\delta(t) + 10^{-6}\frac{d^2v_C}{dt^2} + 4v_C - 4v_o = 0,$$

$$4v_o - 4v_C + \frac{1}{1500}\frac{dv_o}{dt} = 0.$$

Now we need initial values of v_c, v_o, and dv_c / dt. The first two are easy to get, because $v_c(0) = 5$, and $v_o(0) = 1500 \times i_L(0) = 15$ V. The derivative is harder. To find the derivative of a capacitor voltage, we must consider the initial value of the capacitor current because $i_c = C\, dv_c / dt$. Fig. 5.2 shows the necessary currents.

At the top middle node we know $i_L(0)$, which is given. Just a tiny moment after the switch has closed (i.e., at $t = 0^+$), we know the voltage at the left end of the 1-kΩ resistor is 15 V. We also know at this moment the initial capacitor voltage: $v_c(0) = 5$ V. So the voltage across the 1-kΩ resistor is $v_1(0^+) = 15 - 5 = 10$ V.

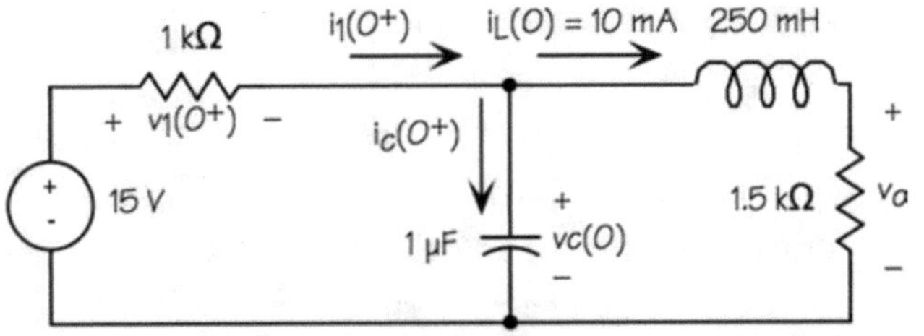

FIGURE 5.2: Initial capacitor current.

From that resistor voltage we get the current arriving from the left: $i_1(0^+) = 10 / 1 = 10$ mA. KCL gives us $i_c(0^+) = i_1(0^+) - i_L(0) = 0$. Hence

$$\left.\frac{dv_C}{dt}\right|_{t=0^+} = \frac{i_c(0^+)}{C} = 0.$$

Now you can use whatever technique you wish to solve the five equations (two differential equations and three initial conditions). The result comes out in nice form (the numbers were rigged!):

$$v_o(t) = (9 + 16e^{-5000t} - 10e^{-2000t})u(t) \text{ V}.$$

Fig. 5.3 is a plot of this output as a function of time.

5.1.2   Solution via State Equations

The "state" of a system is the condition of the energy-storage elements. In other words, the state of a system is related to the energy stored in capacitors and inductors. Recall the energy relationships:

$$w_L(t) = 1/2 \, Li_L^2(t),$$
$$w_C(t) = 1/2 \, Cv_C^2(t).$$

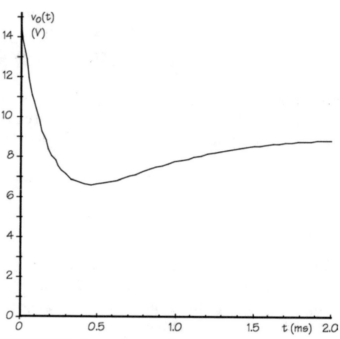

FIGURE 5.3: Output for example.

So the "state" of a circuit can be given by the currents through inductors and the voltages across capacitors. In writing *state equations*, we use these as our variables. If necessary, we add other variables that are related only to nonstorage elements such as resistors.

Let's look at the same circuit that we have just analyzed, repeated in Fig. 5.4.

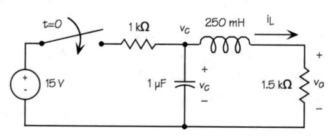

FIGURE 5.4: Example again.

Look first at the top middle node, which I've labeled v_c because its voltage is the capacitor's voltage and hence one of our state variables. I can write a node equation there, using v_c, +15, and i_L:

$$\frac{v_C - 15u(t)}{1000} + 10^{-6}\frac{dv_c}{dt} + i_L = 0.$$

Now look at the current i_L. Note that if I write a single mesh equation around the right-hand mesh, I can use i_L, which the other of our state variables, along with v_c:

$$-v_C + 0.25\frac{di_L}{dt} + 1500i_L = 0.$$

Each equation has only a first derivative and otherwise involves the state variables in the "natural" form. Since we have only first derivatives, we need initial conditions only for those variables. Gee, those are given to us!

$$v_C(0) = 5\text{ V}, i_L(0) = 0.01\text{ A}.$$

These four equations are not hard to solve (and solvers like Maple handle them very readily). It shouldn't surprise you to know that the results are the same as we found before:

$$v_o(t) = (9 + 16e^{-5000t} - 10e^{-2000t})u(t)\text{ V}.$$

Conclusion? State equations require a little more thinking to craft them, but they are generally easier to solve. Moreover, we don't need to find additional initial values because of higher order derivatives.

5.2   ZERO INPUT & ZERO STATE

The zero-input and zero-state concepts are important to our understanding of how circuits function. While they refer to time-domain results, they have an important relationship to the frequency-domain as well. The idea is very simple (dangerous statement?). A circuit has a "starting point." It also has some sources. Our circuits are linear, so we can use superposition. That means we can separate the effects of the "starting points" from the effects of the sources.

So what we are about to do is superimpose two responses: zero input and zero state.

5.2.1   Zero-Input Response

If we have a circuit with energy-storage elements, these energy-storage elements can have stored energy at $t = 0$. This leads to initial conditions. If we now have *only* this initial energy (hence no sources), we will solve and get the *zero-input response*. Note that the words mean exactly what they say: there is no input to the circuit and the only energy comes from the initial stored energy.

Consider our original example, now redrawn in Fig. 5.5. The voltage source on the left has been reduced to a short circuit because its value is zero volts.

I am going to solve this using the original node equations. You'll note that the only change is that the source term has vanished from the first term of the first equation:

$$\frac{v_C}{1000} + 10^{-6}\frac{dv_C}{dt} + \frac{1}{0.25}\int_0^t (v_C - v_o)dx + 10 \times 10^{-3} = 0,$$

$$\frac{1}{0.25}\int_0^t (v_o - v_C)dx - 10 \times 10^{-3} + \frac{v_o}{1500} = 0.$$

The initial values change slightly. Recall that we had to find the initial current through the capacitor in order to find the initial value of the derivative of the capacitor voltage (after differentiating the equations). We need to do this again because the source is now zero.

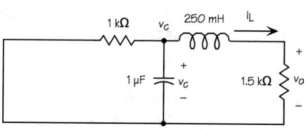

FIGURE 5.5: Example in zero-input form.

Look back at Fig. 5.2 if you need to. The initial capacitor voltage is still $v_c(0) = 5$ V, so the voltage across the 1-kΩ resistor is $v_1(0) = 0 - 5 = -5$ V. Hence the current $i_1(0) = -5 / 1 = -5$ mA. KCL gives us $i_c(0^+) = i_1(0) - i_L(0) = -5 - 10 = -15$ mA. Hence

$$\left.\frac{dv_C}{dt}\right|_{t=0^+} = \frac{i_C(0^+)}{C} = \frac{-15 \times 10^{-3}}{1 \times 10^{-6}} = -15{,}000 \text{ V/s}.$$

The result of solving these equations is the *zero-input* response:

$$v_o(t) = (10e^{-5000t} + 5e^{-2000t})u(t) \text{ V}.$$

Note especially that it dies out! It has to, because when all the initial energy is dissipated, there can't be anything left. Fig. 5.6 is a plot of this part of the output.

5.2.2   Zero-State Response

Conceptually, the zero-state response is a little easier, I think. It just means the initial state of the circuit is zero, which means the initial conditions are zero.

Let's redo the example, this time using the state equations, with the initial conditions set to zero:

$$\frac{v_C - 15u(t)}{1000} + 10^{-6}\frac{dv_c}{dt} + i_L = 0,$$

$$-v_C + 0.25\frac{di_L}{dt} + 1500i_L = 0,$$

$$v_C(0) = 0, \, i_L(0) = 0.$$

The solution of these gives the *zero-state* response (this is plotted in Fig. 5.7):

$$v_o(t) = (9 + 6e^{-5000t} - 15e^{-2000t})u(t) \text{ V}.$$

We can see that this response does not die out (at least in this case). This means that there is still energy in the circuit at the end of time. We should be able to see from both the equation for $v_o(t)$ and its graph that this end-of-time output is 9 V. We recognize this as the *steady-state* response.

Should this steady-state response really be 9 V? Let's look at the circuit at the end of time as shown in Fig. 5.8.

Remember how to figure the end of time? If everything is to be steady (as required by the steady source), we must require certain things of the energy-storage elements:

- An inductor will have a steady current through it (everything has settled down and hence is not changing). If the current is steady, its derivative is zero, so the voltage across such an inductor is zero. The inductor looks like a short circuit, which is what I drew in the upper right branch.

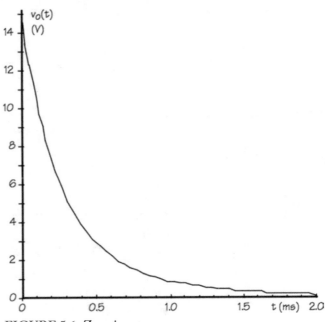

FIGURE 5.6: Zero-input response.

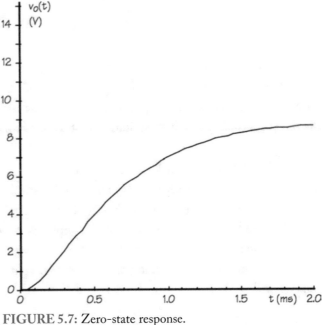

FIGURE 5.7: Zero-state response.

- A capacitor will have a steady voltage across it. By the same reasoning, the derivative of that voltage must be zero, so the current through the capacitor is zero. It looks like an open circuit, which is what I drew in the middle branch.

A simple voltage-divider relationship gives us

$$v_{o\,steady-state} = 15\frac{1.5}{1+1.5} = 9 \text{ V},$$

which is what we just concluded.

5.2.3 Total Response

Since I said that superposition should combine the zero-input response and the zero-state response, perhaps I should try it to show that it works:

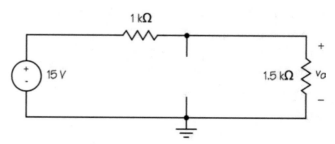

FIGURE 5.8: Example at the end of time.

$$v_o(t) = [(10e^{-5000t} + 5e^{-2000t})$$
$$+(9 + 6e^{-5000t} - 15e^{-2000t})]u(t) \text{ V}$$
$$= (9 + 16e^{-5000t} - 10e^{-2000t})u(t) \text{ V}.$$

Wow! That's what we got twice before. Isn't math grand? But of course, this must be if we believe in superposition in linear systems. Fig. 5.9 shows all three responses. The generally rising line is the zero-state response, the falling line is the zero-input response (which has to die out), and the top line is their sum—or at least it looks to me like their sum.

5.3   TRANSFORM THE EXAMPLE

"Stop!" you say. "We are several pages into this chapter and you haven't done a Laplace transform yet. Why all the buildup and then no transforms?"

I started this chapter with some examples worked in the time domain because that's where we live and that's where we must return after we finish in the frequency domain. In other words, it's important that we know what is happening in the time domain simply because that's "our world."

Since you know from math everything there is to know about the Laplace transform (perhaps even more than you wanted to know), my next step is an easy one. I'm going to take the original time-domain nodal equations and transform them. Then I'll find the solution in the frequency domain (the s domain) and use the inverse Laplace transform to get back to the time domain.

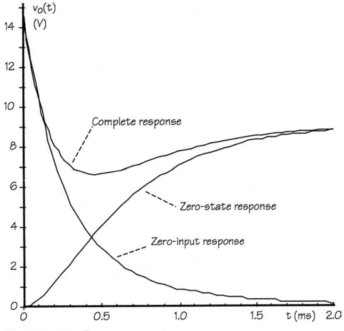

FIGURE 5.9: Response curves.

5.3.1 Example in the Time Domain

I'll repeat here without explanation both the drawing of the circuit from the beginning of the chapter and the time-domain equations that I wrote using nodal analysis. Fig. 5.10 is the circuit as we had it before.

The initial conditions are

$$i_L(0) = 10 \text{ mA},$$
$$v_C(0) = 5 \text{ V}.$$

The node equations that I wrote before are

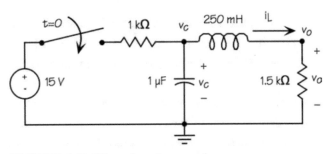

FIGURE 5.10: Time-domain example.

$$\frac{v_C - 15u(t)}{1000} + 10^{-6}\frac{dv_C}{dt} +$$

$$\frac{1}{0.25}\int_0^t (v_C - v_o)dx + 10 \times 10^{-3} = 0,$$

$$\frac{1}{0.25}\int_0^t (v_o - v_C)dx - 10 \times 10^{-3} + \frac{v_o}{1500} = 0.$$

While I am at it, let's recall the time-domain solution, too:

$$v_o(t) = (9 + 16e^{-5000t} - 10e^{-2000t})u(t) \text{ V}.$$

Keep in mind that this work has all been done in the *time domain*, which means that t is the independent variable. Now let's switch to the frequency domain, where the independent variable will be s, whose units are *rad/s*.

5.3.2 Transforming the Equations

Use the Laplace transform to transform these equations into the frequency domain. The result of a straightforward application of the transform gives

$$\frac{V_C(s)}{1000} - \frac{15}{1000s} + 10^{-6}(sV_C(s) - v_C(0)) +$$

$$\frac{1}{0.25}\frac{V_C(s) - V_o(s)}{s} + \frac{0.01}{s} = 0,$$

$$\frac{1}{0.25}\frac{V_o(s) - V_C(s)}{s} - \frac{0.01}{s} + \frac{V_o(s)}{1500} = 0,$$

$$v_C(0) = 5 \text{ V}.$$

The initial conditions found their way very simply into our transformed equations. The initial value of the inductor current was already part of the time-domain equations; it appears divided by s. The initial capacitor voltage is introduced into the transformed equations when we take the Laplace transform of the derivative of that voltage.

This is a nice advantage of using the Laplace transform—the initial conditions "come along for free." This means we don't have to take special pains to determine them, provided we know them in terms of capacitor voltages and inductor currents.

Also recall that the Laplace transform starts at $t = 0$, so we must have starting values for our variables at that point. Sometimes we say that we really want the Laplace transform to start *just before* $t = 0$, or said another way, to start at $t = 0^-$. This is important only if there is an impulse at the origin. That impulse, which represents an initial blast of energy into the circuit, exists only mathematically, however. They don't happen in the "real" world.

5.3.3 Solving in the s-Domain

We have the equations in the frequency domain, where s is the independent variable. Now let's solve them. After doing this, and using some algebra to get them into a "nice" form, we have

$$V_o(s) = 15\frac{s^2 + 3000s + 6 \times 10^6}{s(s^2 + 7000s + 10^7)}.$$

It's interesting to factor the denominator to see what the individual factors look like:

$$V_o(s) = 15 \frac{s^2 + 3000s + 6 \times 10^6}{s(s + 5000)(s + 2000)}.$$

Hmmm, we can see the individual parts of the time-domain response sitting right there in the denominator (recall partial fractions):

- The single s, when taken back to the time domain, is a step function.
- The term $s + 5000$ will become an exponential of the form e^{-5000t}.
- The term $s + 2000$ also will be an exponential of the form e^{-2000t}.

If we remember the time-domain solution, we recognize that this frequency-domain response contains the same information. We haven't lost anything by this transformation—we've just shifted to another domain.

How about the units of $V_o(s)$? If you track units carefully, you'll recognize that the "15" in the response is the voltage of the step input. Hence it has the units of *volts*. The numerator is in s^2, so it has the units of $(rad/s)^2$, while the denominator, in s^3, has the units of $(rad/s)^3$.

Combining all these units gives us *volt-seconds* as the unit for $V_o(s)$. But now that we know that, we'll probably do like almost everyone else and get sloppy. We will *say* the units are volts, and we'll try to remember that they really aren't! Amperes should be *ampere-seconds* as well.

One more thing: the partial fraction expansion of $V_o(s)$? Such an expansion shows us clearly the individual parts of the solution:

$$V_o(s) = \frac{9}{s} + \frac{16}{s + 5000} - \frac{10}{s + 2000}.$$

In fact, this result is so simple that you may be able to return to the time domain by inspection. Using either that method or the inverse Laplace transform, we get the result we have seen before:

$$v_o(t) = (9 + 16e^{-5000t} - 10e^{-2000t})u(t) \text{ V}.$$

There! We are back to the time domain, no worse for the trip, I hope. Uh, what now? You say you could have solved this problem just as easily without all this Laplace stuff? Heresy, of course! But you are probably right. In linear systems, and especially with a math-doer like Maple, the Laplace transform adds some complexity.

But don't go away, because we can enter the frequency domain directly from the circuit itself.

5.4   TRANSFORMING THE CIRCUIT

So far, we've solved a problem entirely in the time domain, where t is the independent variable. Then we've solved it again, this time converting the equations, using the Laplace transform, to the frequency domain. There, the independent variable is s, a frequency in radians per second.

Now for some good news—we can skip the time domain entirely if we want to! The circuit itself can be drawn in the s-domain and then analyzed entirely in the s-domain. In this section we'll learn how to do that.

We have been working with linear systems. The Laplace transform is a linear transformation. So whether we apply it before or after writing the circuit equations makes no difference.

But something quite neat happens. When we write equations for our circuit in the time domain, we get differential equations. When we do this in the frequency domain, we get algebraic equations. There is some advantage to that, I would say.

So let's see how to transform the circuit.

5.4.1   Transforming the Elements

Consider transforming sources. They are easy to do. Just take the Laplace transform of the time function. There's one small point that leads to errors, though. Suppose a source is 5 V DC. The transform of this *must be* 5/s. In other words, we must treat the constant source as if it were a step function—it acts like it starts at $t = 0$.

Now consider the circuit elements. I'll start with the resistor. In the time domain, we write

$$v(t) = Ri(t).$$

When we transform this, $v(t)$ and $i(t)$ are generically transformed into $V(s)$ and $I(s)$. The result is

$$V(s) = RI(s).$$

This says the resistor R in the time domain looks just like a resistor R in the frequency domain as shown in Fig. 5.11.

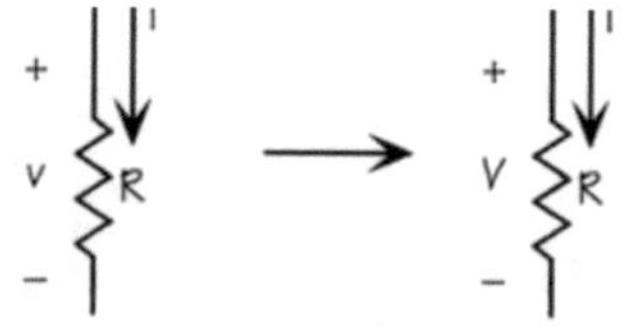

Next let's do the inductor, which is described with a derivative:

$$v(t) = L\frac{di}{dt}.$$

FIGURE 5.11: Transforming a resistor.

Transforming this involves an s, but more important, it brings in the initial value of the current through the inductor. In other words, the transform of the inductor includes its initial *state:*

$$V(s) = L[sI(s) - i(0)] = LsI(s) - Li(0).$$

The inductor in the frequency domain requires a bit more complicated model than did the resistor. $V(s)$ is composed of two terms, both voltages. So the model will have the current $I(s)$ passing through two elements in series, one for the term $LsI(s)$ and one for the term $-Li(0)$.

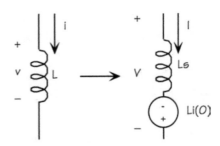

FIGURE 5.12: Transforming an inductor.

The first element of our model will be shown as an inductor whose frequency-domain value is Ls. The second will be a voltage source whose value is $-Li(0)$. Fig. 5.12 shows this circuit.

What are the units of the Ls term? Remember when you were learning about impedances? Inductors became $j\omega L$, which had the unit of ohms. Well, s is a frequency in radians per second just as ω is. So the unit for Ls is likewise ohms, and we say that Ls is the impedance of the inductor.

We can also model the inductor in another way. If we reverse the time-domain equation for an inductor, we get

$$i(t) = \frac{1}{L} \int_0^t v(x)dx + i(0).$$

Here I have included the initial value of the current since the integral starts at $t = 0$. The transform of this is

$$I(s) = \frac{V(s)}{Ls} + \frac{i(0)}{s}.$$

Gee, two terms again. Both are currents that sum to become $I(s)$. So these must be two elements in parallel with a common voltage $V(s)$ across them. One element represents the inductor, whose impedance is Ls. The second is a current source whose value is $i(0)/s$. Here again, the initial current appears.

Fig. 5.13 shows both transformations of the inductor.

Finally comes the capacitor. Let's transform it with a little less talk. The time-domain equation and its transform into the frequency domain are

$$v(t) = \frac{1}{C} \int_0^t i(x)dx + v(0),$$

$$V(s) = \frac{1}{Cs} I(s) + \frac{v(0)}{s}.$$

The frequency-domain model of the capacitor will be the series combination of the impedance of the capacitor ($1/Cs$) and a term representing the initial value of the capacitor voltage ($v(0)/s$).

We can also write the time-domain equation for the capacitor in the "other" order and thereby get a different transform:

$$i(t) = C\frac{dv}{dt},$$
$$I(s) = C[sV(s) - v(0)] = CsV(s) - Cv(0).$$

This model will be the parallel combination of the impedance of a capacitor ($1/Cs$) and a current source representing the initial capacitor voltage.

Fig. 5.14 shows these transformations.

These models simplify when we are looking for the zero-state response. All the initial-condition sources can be eliminated. That means resistors are R, inductors are Ls, and capacitors are $1/Cs$ ohms.

5.4.2 Transforming a Circuit

Here's an example of transforming the circuit itself. Start with the circuit shown in Fig. 5.15. Presume there is energy stored in each capacitor and inductor at $t = 0$, represented by $i_L(0)$, $v_{C1}(0)$, and $v_{C2}(0)$.

If we plan to write equations using nodal analysis, we will find it easier to use the current-source forms of the models. In Fig. 5.16, the source becomes v_i/s, the inductor is Ls in parallel with the transformed initial condition, and each capacitor is $1/Cs$ in parallel with its transformed initial condition. Resistors are the same. (Also note that the output voltage v_o (small letter) became V_o (capital letter).)

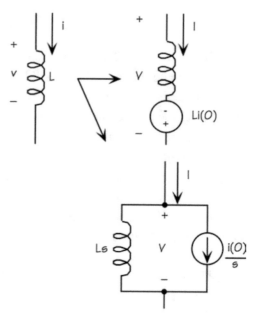

FIGURE 5.13: Transforming an inductor.

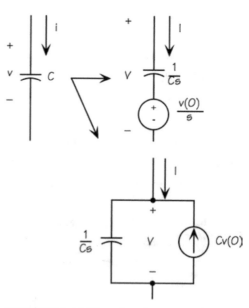

FIGURE 5.14: Transforming a capacitor.

If we plan to write mesh equations, the voltage-source models make the equations simpler. Fig. 5.17 shows the same example but with voltage sources.

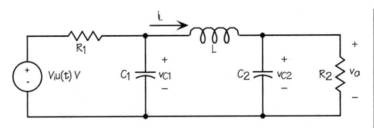

FIGURE 5.15: Circuit to transform.

5.4.3 Transformed Circuit with Numbers

How about another example, this time with numbers? The circuit is shown in Fig. 5.18.

Initial conditions are

$$i_L(0) = 0, v_C(0) = -2 \text{ V}.$$

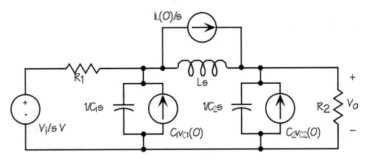

FIGURE 5.16: Transformed circuit (for nodes).

Transforming this for nodal analysis gives the circuit shown in Fig. 5.19. Note that the sine wave input is also transformed.

If we instead want to use mesh analysis, we would transform the circuit as shown in Fig. 5.20.

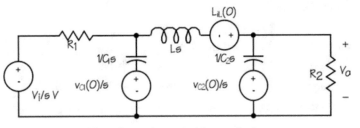

FIGURE 5.17: Transformed circuit (for meshes).

5.5 ORIGINAL EXAMPLE TO *S*

Can you stand looking once more at the circuit that opened this chapter? Try hard! Fig. 5.21 repeats the original drawing in the time domain.

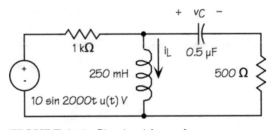

FIGURE 5.18: Circuit with numbers.

The initial conditions are

$$i_L(0) = 10 \text{ mA},$$
$$v_C(0) = 5 \text{ V}.$$

5.5.1 s-Domain Solution

I will use nodal analysis, so I need to use the current-source models of the inductor and the capacitor. The source is a constant voltage with a switch. This is modeled as a step function and hence transforms as $15/s$. The transformed circuit is shown in Fig. 5.22.

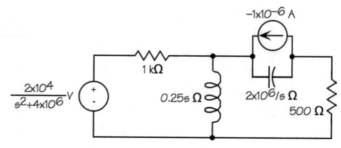

FIGURE 5.19: Circuit transformed (for nodes).

Now for node equations:

$$\frac{V_C - 15/s}{1000} + \frac{V_C}{10^6/s} - 5 \times 10^{-6} + \frac{0.01}{s} + \frac{V_C - V_o}{0.25s} = 0,$$

$$\frac{V_o - V_C}{0.25s} - \frac{0.01}{s} + \frac{V_o}{1500} = 0.$$

Note that these are algebraic and that the initial conditions are automatically included in them.

The solutions in the frequency domain and in the time domain are the same as before.

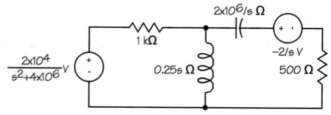

FIGURE 5.20: Circuit transformed (for meshes).

$$V_o(s) = 15 \frac{s^2 + 3000s + 6 \times 10^6}{s(s + 5000)(s + 2000)},$$

$$= \frac{9}{s} + \frac{16}{s + 5000} - \frac{10}{s + 2000},$$

$$v_o(t) = (9 + 16e^{-5000t} - 10e^{-2000t})u(t) \text{ V}.$$

5.5.2 Zero Input and Zero State

The zero-input (no sources) response can be gotten by replacing the original voltage source with a short circuit (for zero voltage). The sources in our transformed circuit that represent initial conditions must remain. Fig. 5.23 is the transformed circuit with the original source dead.

Writing node equations yields

$$\frac{V_C}{1000} + \frac{V_C}{10^6/s} - 5 \times 10^{-6} + \frac{0.01}{s} + \frac{V_C - V_o}{0.25s} = 0,$$

$$\frac{V_o - V_C}{0.25s} - \frac{0.01}{s} + \frac{V_o}{1500} = 0.$$

That's not much different from before! Only the numerator of the first term has changed to eliminate the 15-V source.

The solution is

$$V_o(s) = 15\frac{s + 3000}{(s + 5000)(s + 2000)},$$

$$= \frac{10}{s + 5000} + \frac{5}{s + 2000},$$

$$v_o(t) = (10e^{-5000t} + 5e^{-2000t})u(t)\text{ V},$$

which is what we got before.

The zero-state response is found by eliminating all the sources that provide initial conditions. Fig. 5.24 shows this arrangement.

Node equations aren't much different this time, either:

$$\frac{V_C - 15/s}{1000} + \frac{V_C}{10^6/s} + \frac{V_C - V_o}{0.25s} = 0,$$

$$\frac{V_o - V_C}{0.25s} + \frac{V_o}{1500} = 0.$$

The zero-state solution is

$$V_o(s) = \frac{9 \times 10^7}{s(s + 5000)(s + 2000)},$$

$$= \frac{9}{s} + \frac{6}{s + 5000} - \frac{15}{s + 2000},$$

$$v_o(t) = (9 + 6e^{-5000t} - 15e^{-2000t})u(t)\text{ V}.$$

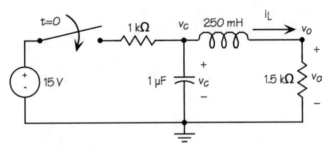

FIGURE 5.21: Original example for the last time.

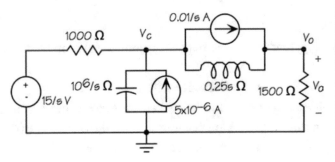

FIGURE 5.22: Transformed example.

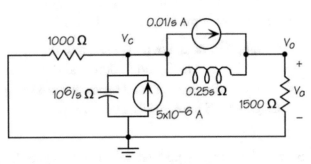

FIGURE 5.23: Zero-input circuit.

Again, it's what we got before.

Do the zero-input and zero-state solutions sum to the total solution in the frequency domain? Well, our circuit is linear, the Laplace transformation is linear, and superposition should work. Let's try it:

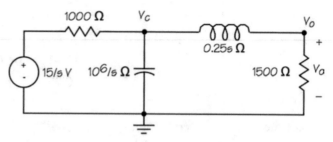

FIGURE 5.24: Zero-state circuit.

$$V_{o\,total}(s) = V_{o\,zero-input}(s) + V_{o\,zero-state}(s)$$

$$= \frac{9}{s} + \frac{6}{s+5000} - \frac{15}{s+2000}$$

$$+ \frac{10}{s+5000} + \frac{5}{s+2000}$$

$$= \frac{9}{s} + \frac{16}{s+5000} - \frac{10}{s+2000}.$$

Yup, that's the same frequency-domain solution we had before.

5.5.3 Steady-State Result

Recall that our example has a steady-state solution that we checked way back at the beginning of this chapter. A long time after $t = 0$, the output is $v_o(t) = 9$ V. This is after all the time-varying functions (here, exponentials) have died out.

This steady-state solution is visible in the frequency domain, too. Consider the total solution written in partial-fraction form:

$$V_o(s) = \frac{9}{s} + \frac{16}{s+5000} - \frac{10}{s+2000}.$$

There are three distinct terms. Each will, through the inverse Laplace transform, become a simple function of time in the time domain:

- The first term, $9/s$, transforms into the step function $9\,u(t)$, which does not die out.
- The second term, $16/(s+5000)$, transforms into the exponential $16\,e^{-5000t}\,u(t)$, which has a negative exponent and hence dies out as t gets large.
- The third term, $-15/(s+2000)$, transforms into the term $-15\,e^{-2000t}\,u(t)$, which also dies out.

So the first term is all that remains as t gets large, which means that the steady-state solution is

$$V_{o\,steady-state}(s) = \frac{9}{s}, v_{o\,steady-state}(t) = 9 \text{ V}.$$

5.5.4   Initial and Final Value

Another thing you might have learned about Laplace transforms in math is the initial-value and final-value theorems.

The initial-value theorem is

$$\lim_{t \to 0+} f(t) = \lim_{s \to \infty} sF(s).$$

This tells us that we can find, from the Laplace transform, the value of f(t) at t = 0. (If there is a singularity at the origin, we get the value of f(t) just barely to the right past t = 0.)

There is a restriction on the initial-value theorem. F(s) must be a proper rational function, which means the numerator must be at least one degree less than the denominator.

The final-value theorem is

$$\lim_{t \to \infty} f(t) = \lim_{s \to 0} sF(s).$$

This tells us that we can find the steady-state response, which is the response as t goes to infinity, from the Laplace transform. There is one restriction here, too. The denominator of $sF(s)$ must have roots whose real parts are negative. This is easy to check: there must be no minus signs on any terms in the denominator (or equivalently, *all* signs must be minus) and no power of s is skipped.

(We will learn later that the roots of the denominator of $F(s)$ must lie in the left half of the complex plane.)

What good are these in circuit analysis? Well, there's the obvious statement that these allow us to find easily the value of $f(t)$ at the beginning or the end of time.

But there's another equally important "good." These allow us to make a fairly simple check on our work. After all, we'll be manipulating algebraic equations and they can get messy. A simple check provided by these theorems can uncover gross errors quickly.

Let's try these on the solution to $v_o(t)$ in our running example. Recall that the solution was

$$V_o(s) = 15 \frac{s^2 + 3000s + 6 \times 10^6}{s(s + 5000)(s + 2000)},$$

$$v_o(t) = (9 + 16e^{-5000t} - 10e^{-2000t})u(t) \text{ V}.$$

Try the initial-value theorem on $V_o(s)$:

$$\lim_{t \to 0} v_o(t) = \lim_{s \to \infty} sV_o(s)$$

$$= \lim_{s \to \infty} 15s \frac{s^2 + 3000s + 6 \times 10^6}{s(s + 5000)(s + 2000)}$$

$$= \lim_{s \to \infty} 15s \frac{s^2}{s^3} = \lim_{s \to \infty} 15 \frac{s^3}{s^3} = 15 \text{ V}.$$

Sure enough, if we look at $v_o(t)$ for $t = 0$, the result is $9 + 16 - 10 = 15$ V.

Now try the final-value theorem:

$$\lim_{t \to \infty} v_o(t) = \lim_{s \to 0} sV_o(s)$$

$$= \lim_{s \to 0} 15s \frac{s^2 + 3000s + 6 \times 10^6}{s(s + 5000)(s + 2000)}$$

$$= \lim_{s \to 0} 15s \frac{6 \times 10^6}{s(10^7)} = \lim_{s \to 0} 15 \frac{6 \times 10^6}{10^7} = 9 \text{ V}.$$

Wow! That worked, too.

But watch out for the restrictions! These theorems will give proper-looking results, even when the restrictions are violated. But the results will most likely be wrong.

5.6 S'MORE EXAMPLES

These examples will take you through writing equations for a circuit in several forms: nodal, state, zero input, zero state, and s-domain. There is one example of initial and final values and then an s-domain design problem.

5.6.1 Example I

For $t > 0$, find the voltage across the capacitor $v_C(t)$ in the circuit of Fig. 5.25. The switch has been open for a long time before $t = 0$. It closes at $t = 0$ to short out one of the 50-Ω resistors.

Since there has been a source attached to the circuit for a long time before $t = 0$, there is some initial energy storage. Since the voltage across the capacitor is steady before $t = 0$, the current through it just before $t = 0$ is 0. Therefore the current through the inductor is caused by the 30-V source

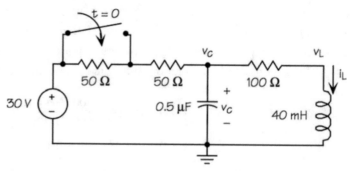

FIGURE 5.25: Examples I and II.

across all the resistors in series. The current through the inductor at $t = 0$ is steady, so the voltage across it just before $t = 0$ is 0. Hence the voltage across the capacitor can be found from that current passing through the 100-Ω resistor.

$$i_L(0) = \frac{30}{50 + 50 + 100} = 150\,\text{mA},$$
$$v_C(0) = 100 i_L(0) = 15\,\text{V}.$$

I'll write node equations with a reference node at the bottom and node voltages as labeled. This yields one differential equation and one integral equation:

$$\frac{v_C - 30}{50} + 0.5 \times 10^{-6} \frac{dv_C}{dt} + \frac{v_C - v_L}{100} = 0,$$
$$\frac{v_L - v_C}{100} + \frac{1}{40 \times 10^{-3}} \int_0^t v_L \, dx + i_L(0) = 0.$$

Solving these gives

$$v_C(t) = 20 + 3.078 e^{-3885t} - 8.078 e^{-38615t}\,\text{V for t} > 0.$$

Note that, after the exponentials die out, the capacitor voltage settles down to 20 V, which is what a voltage divider, applied to the circuit after a long time after $t = 0$, gives us.

5.6.2   Example II

Redo the previous example by writing state equations.

State equations must include the "state variable" for each energy-storage element. In this example, the capacitor voltage $v_C(t)$ and the inductor current $i_L(t)$ represent the state of the system.

My equations come from a node equation at the top of the capacitor and a mesh equation around the mesh on the right:

$$\frac{v_C - 30}{50} + 0.5 \times 10^{-6} \frac{dv_C}{dt} + \frac{v_C - v_L}{100} = 0,$$
$$-v_C + 100 i_L + v_L = 0,$$
$$v_L = 40 \times 10^{-3} \frac{di_L}{dt}.$$

When the initial conditions as calculated previously are combined with these equations, the result comes out the same. (It would have been very hard to explain if the result were different!)

5.6.3 Example III

Redo the previous example, but for zero-input operation.

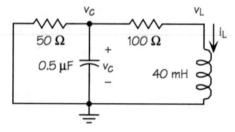

The initial values are the same as before because the source was present until $t = 0$. After $t = 0$, the circuit is as shown in Fig. 5.26.

FIGURE 5.26: Example III after switching.

The equations come out looking almost the same. I used the state equations we just had in the previous section and removed the 30-V source term:

$$\frac{v_C}{50} + 0.5 \times 10^{-6} \frac{dv_C}{dt} + \frac{v_C - v_L}{100} = 0,$$

$$-v_C + 100 i_L + v_L = 0,$$

$$v_L = 40 \times 10^{-3} \frac{di_L}{dt}.$$

The zero-input result is

$$v_{C-0input}(t) = -9.236 e^{-3885t} + 24.236 e^{-38615t} \text{ V for } t > 0.$$

5.6.4 Example IV

Now analyze the same circuit but this time under the zero-state condition.

This requires changing things a little as shown in Fig. 5.27. The switch/50-Ω combination is replaced by a switch that connects the source to the circuit at $t = 0$. Now the initial values are both zero.

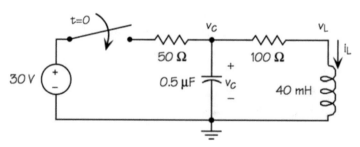

FIGURE 5.27: Example IV.

The state equations are the same, but the initial conditions change:

$$v_C(0) = 0, i_L(0) = 0.$$

The zero-state result is

$$v_{C-0\,state}(t) = 20 + 12.313e^{-3885t} - 32.313e^{-38615t} \text{ V for } t > 0.$$

The zero-state and zero-input results should add (superimpose!) to become the original solution. They do, and this is a good check on my work.

5.6.5   Example V

One more time! Analyze the same circuit by converting it first to the s-domain and then writing and solving algebraic equations.

Fig. 5.28 shows the converted circuit.

Note the changes:

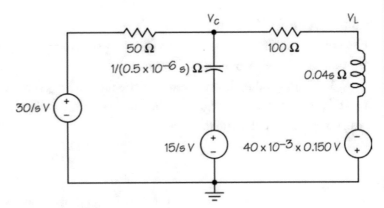

FIGURE 5.28: Example V converted to s-domain.

- The source must be taken as a step function because the transformed version of the circuit doesn't exist before $t = 0$. Hence it becomes $30/s$ V. (Well, technically volt-seconds, but remember that most of us get sloppy and forget the "seconds" part of the units in the s-domain.)
- The switch is gone because it is closed from $t = 0$ on.
- The capacitor is $1/Cs$ and the inductor is Ls.
- For both initial conditions I have chosen to use voltage source models.
- The initial capacitor voltage becomes $v_C(0)/s$.
- The initial inductor current becomes $Li_L(0)$. Note that this source is "upside down."

My node equations, using the variables as marked on the circuit, are

$$\frac{V_C - 30/s}{50} + \frac{V_C - 15/s}{1/0.5 \times 10^{-6}} + \frac{V_C - V_L}{100} = 0,$$

$$\frac{V_L - V_C}{100} + \frac{V_L - (-40 \times 10^{-3} \times 0.150)}{0.04s} = 0.$$

Solving these algebraically gives the s-domain result

$$V_C(s) = 15 \frac{s^2 + 62500s + 200 \times 10^6}{s\left(s^2 + 42500s + 150 \times 10^6\right)} \text{ V.}$$

This is given in one of the "standard" forms: the ratio of two polynomials in s with 1 as the coefficient of the highest power of s in both the numerator and the denominator.

Using the inverse Laplace transform to find $v_C(t)$ yields the same result as in Examples I and II.

5.6.6   Example VI

The initial- and final-value theorems should give results, calculated in the s-domain, that are the same as what we would get in the time domain.

The initial-value theorem requires the function $V_C(s)$ to be a proper rational fraction. It is in this case because the degree of the numerator is less than the degree of the denominator. I can apply the initial-value theorem:

$$v_C\left(0^+\right) = \lim_{s \to \infty}\left(sV_C(s)\right)$$
$$= \lim_{s \to \infty}\left[s\left(15\frac{s^2 + 62500s + 200 \times 10^6}{s\left(s^2 + 42500s + 150 \times 10^6\right)}\right)\right]$$
$$= 15 \text{ V.}$$

This is the same result as the initial condition for the capacitor voltage that we found in Example I.

The final-value theorem requires that all the roots of the denominator have negative real parts. In this case, no power of s is skipped and all coefficients are positive, so I can apply the final-value theorem:

$$v_C\left(\infty\right) = \lim_{s \to 0}\left(sV_C(s)\right)$$
$$= \lim_{s \to 0}\left[s\left(15\frac{s^2 + 62500s + 200 \times 10^6}{s\left(s^2 + 42500s + 150 \times 10^6\right)}\right)\right]$$
$$= 20 \text{ V.}$$

This is the same value that we get from Example I when we let t become very large.

5.6.7   Example VII

Design a circuit that has a voltage transfer function $H(s) = V_{out}(s) / V_{in}(s) = 1000 / (s + 1500)$.

Design problems like this can be done by drawing generic circuits until you get one that has the "numbers in the right place." But it's possible to use your head and draw some conclusions before drawing a circuit:

- At DC ($s = 0$), the voltage transfer function is 2/3, which implies a voltage divider consisting of two resistors.
- There is a single s term in the denominator, so there is only one energy-storage element.
- I don't like inductors!
- Paralleling the output resistor with a capacitor might work.

My generic attempt is shown in Fig. 5.29. At DC, the output is the input times $R_2/(R_1 + R_2)$. The capacitor will contribute an s somewhere.

I wrote and simplified the generic equation for the voltage transfer function:

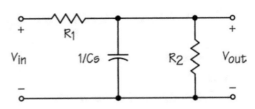

FIGURE 5.29: Example VII: possible circuit.

$$H(s) = \frac{R_2 \left\| \dfrac{1}{Cs} \right.}{R_2 \left\| \dfrac{1}{Cs} \right. + R_1} = \frac{R_2}{R_2 + R_1 R_2 Cs + R_1}$$

$$= \frac{\dfrac{1}{R_1 C}}{s + \dfrac{1}{R_1 C} + \dfrac{1}{R_2 C}}.$$

Hmmm, that looks pretty good. I'll match up some of the terms:

$$\frac{1}{R_1 C} = 1000, \frac{1}{R_1 C} + \frac{1}{R_2 C} = 1500.$$

Then I'll reduce these so that I have the resistors in terms of the capacitor C. I do this because there aren't a lot of different commercial values of capacitance and lots of resistance values.

$$R_1 = \frac{1}{1000C}, R_2 = \frac{1}{500C}.$$

I chose $C = 1\ \mu$F. This choice made $R_1 = 1\ k\Omega$ and $R_2 = 2\ k\Omega$. Those values sound "pretty good." That means they are sensible values for a typical circuit carrying signals. In other words, they are in the "kilo" range. Fig. 5.30 is my final circuit.

If I hadn't been happy with these elements, I could have chosen a different value for C. In fact, I might want to choose a smaller value because 1μF is somewhat large. If I do that, the resistance values will be larger, which will be OK, too.

5.7   CIRCUIT DESIGN EXAMPLE

The circuit shown as a box in Fig. 5.31 is to have a zero-state frequency-domain response of

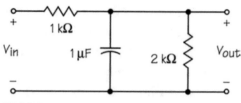

$$H(s) = \frac{V_o(s)}{V_{in}(s)} = \frac{0.1}{(1+s/1000)(1+s/10,000)}.$$

FIGURE 5.30: Example VII: one result.

Design a possible circuit to go into the box.

Hmmm, what's going to work here? Let's wander through some thoughts:

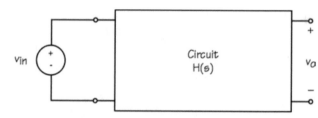

FIGURE 5.31: Circuit design form.

- There's an s^2 in the denominator, whose units are $(rad/s)^2$.
- As s increases, $H(s)$ decreases, so we want a circuit that reduces the output as s increases.
- Capacitors behave like $1/Cs$, so capacitors might be a good choice, connected across the output.
- To have an s^2 in the denominator, we'll need two capacitors.
- To make the capacitors act somewhat separately, let's put a resistor in series between them.
- To separate the capacitors from the source, let's also put a resistor in series with the input.

The circuit of Fig. 5.32 has the characteristics I've just thought about.

I've already put the circuit into the frequency domain. In fact, all my thinking has been in the frequency domain because, after all, that's how the problem was stated.

The next step is to see if we can make anything work here, so I'll write node equations and solve them. In the equations that follow, V_1 is the voltage at the top of C_1, and the reference node is at the bottom:

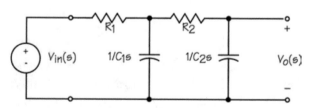

FIGURE 5.32: Possible circuit.

$$\frac{V_1 - V_{in}}{R_1} + \frac{V_1}{1/C_1 s} + \frac{V_1 - V_o}{R_2} = 0,$$

$$\frac{V_o - V_1}{R_2} + \frac{V_o}{1/C_2 s} = 0.$$

The solution is

$$\frac{V_o}{V_{in}} = \frac{1}{1 + (C_1 R_1 + C_2 R_1 + C_2 R_2)s + C_1 C_2 R_1 R_2 s^2} = H(s).$$

This must match the original specification for $H(s)$, which I have expanded to look like the equation above:

$$H(s) = \frac{0.1}{1 + 11 \times 10^{-4} s + 10^{-7} s^2}.$$

This match can be satisfied by

$$C_1 C_2 R_1 R_2 = 10^{-7},$$

$$C_1 R_1 + C_2 R_1 + C_2 R_2 = 11 \times 10^{-4}.$$

Hmmm, two equations and four unknowns. That's not an unlikely happening when you are designing a circuit. So now I must make some judicious choices. I think I will make the two capacitors equal, and I'll choose a reasonable, commercially available value of 1 µF. Then I can solve these equations for R_1 and R_2.

Why choose the capacitor values? There are very few standard values of commercial capacitors, while there are many values of 5% resistors. Hence I am more likely to find resistors that fit than capacitors. Here's the solution for the resistances. There are two possible solutions.

$$C_1 R_1 + C_2 R_1 + C_2 R_2 = 11 \times 10^{-4}.$$

$$C_1 = C_2 = 1\,\mu F,$$

$$R_1 R_2 = \frac{10^{-7}}{10^{-6} + 10^{-6}} = 10^5 \text{ (OK!)},$$

$$10^{-6} R_1 + 10^{-6} R_1 + 10^{-6} R_2 = 11 \times 10^{-4},$$

$$2R_1 + R_2 = 1100 \text{ (also OK!)},$$

$$R_1 = 435.1\,\Omega, R_2 = 229.8\,\Omega,$$

$$\text{or } R_1 = 114.9\,\Omega, R_2 = 870.2\,\Omega.$$

I am going to choose the first set of results, giving R_1 = 435.1 Ω and R_2 = 229.8 Ω. Plugging these values back into the $H(s)$ for the circuit will show that this matches the desired $H(s)$—almost!

$$H(s) = \frac{1}{1+11\times 10^{-4} s + 10^{-7} s^2}.$$

Note that the numerator is 1 instead of 0.1. Oh, I think, just hang a voltage divider on the output. Well, that will work, but I must choose values that will not "load" the output of the RC circuit enough to alter its characteristics.

Instead, I am going to use an op-amp to fix the level. This is often a good solution if the frequencies involved are not too high and the signal levels are in the range an op-amp can handle.

While I would like to use a noninverting op-amp circuit (because the input resistance to the plus terminal is so very high), I can't. This circuit needs a gain of 0.1, and a noninverter can't provide less than 1 (at least not without some modification of where the output is connected).

I chose an input resistor of 100 kΩ because that will have a very small effect on the preceding circuit. The op-amp has a gain of -0.1, which should be acceptable if inversion of the signal is OK (it often is).

But the resistors I've chosen aren't commercial values. I could use more expensive 1% resistors, the nearest of which are R_1 = 437 Ω, R_2 = 229 Ω. But use of 1% resistors probably isn't economically justified. So I'll use 5% values: R_1 = 430 Ω, R_2 = 220 Ω.

How does this check out? Using those 5% values,

$$H(s) = \frac{0.1}{1+10.8\times 10^{-4} s + 0.946\times 10^{-7} s^2}$$
$$= \frac{0.1}{(1+s/1016.4)(1+s/10400)}.$$

The roots are out of position by at worst 4%. That's probably acceptable. The resultant circuit is shown in Fig. 5.33.

5.8   SUMMARY

We have looked at how the Laplace transform can be applied in circuit analysis by working from time-domain equations to transformed equations to transformed circuits.

Now is a good time to see graphically what we've been doing. Look at Fig. 5.34, which shows the different solution paths that we've followed.

On the left are the events in the time domain. We begin with the circuit, write equations in the time domain,

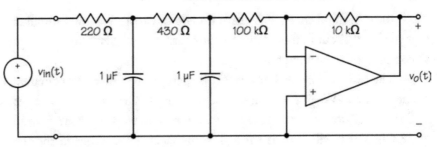

FIGURE 5.33: One design for $H(s)$.

equations that are differential if there are any energy-storage elements. Then we solve these equations using techniques we learned in a class on differential equations. The result is a solution to our circuit problem, stated in the time domain.

The dotted lines show the path we took to solve the time-domain equations in the frequency domain. We started with the circuit and wrote the equations. Then we transformed these equations into the frequency domain. The Laplace transform carried with it the initial conditions. We solved the transformed equations, which were algebraic in nature. That gave us a solution for the circuit problem, but stated in the frequency domain. The inverse Laplace transform brought us back to the time domain.

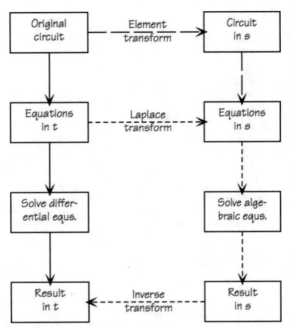

FIGURE 5.34: Solution paths.

The dashed lines at the top show a third path from circuit to solution. This time we transformed the circuit itself into the frequency domain, generating a circuit that has elements as functions of s. Then we wrote and solved the algebraic equations to get the frequency-domain solution. As before, if we wanted to, we went back to the time domain.

What's important out of this? Two things. We need to be able to follow either of the two outside paths in Fig. 5.34. In other words, we must be able to solve problems either in the time domain or in the frequency domain. Being able to transform the circuit from the time domain and back to it is an important part of this.

Keep in mind that the Laplace transform carries lots of manipulative baggage, baggage that you learn about in math courses. Our basic use of this transform is pretty straightforward, so computers can be a big help.

The rest of *Pragmatic Circuits* II is going to be devoted to circuits primarily in the frequency domain. We like to analyze and also design circuits in this frequency domain, because, as Fig. 5.34 shows, we can use algebra. And we can use it in some powerful ways.

Chapter 6 will take us through some facets of working in the frequency domain with functions of s. Then in Chapter 7 we'll restrict that look to just the $j\omega$ part of s. Finally, in Chapter 8 we will restrict that frequency to 60 Hz as we look at AC power circuits.

CHAPTER 6

Frequency Domain: Where S is King

Think back into the dim past, all the way back to Chapter 5 that is, and recall that we entered the frequency domain. This is nothing more, at least in math, than changing the word following "f of…" when we speak about functions. Instead of "f of t" for the time domain, we are now saying "F of s" for the frequency domain.

While the time domain is important (I kinda like to think I live in the time domain), the frequency domain is important in electrical systems. Not only is it a nice convenience and work saver, making differential equations into algebraic ones, it is also a very good way to represent some important properties and behaviors of electrical systems.

In this chapter we are going to take a closer look at three subjects in the frequency domain:

- the impedance $Z(s)$,
- the transfer function $H(s)$, and
- the s-plane representation of these.

Before we do these, though, let's think back to Chapter 5. Recall that we had circuit responses that we categorized as either

- *zero state*, which means that the circuit has no initial stored energy (i.e., capacitor voltages and inductor currents at $t = 0$ are all zero); and
- *zero input*, which means there are no sources driving the circuit, so that whatever the circuit does, it does it because of initial stored energy.

Everything we are going to do in this chapter is based on *zero-state* conditions. There is *no* initial stored energy. If there is, our definitions will begin to fall apart. Keep this in mind as we work through the examples.

Now think back to Chapter 2 where we took up the formal methods of nodal and mesh analysis. And think back to Chapter 3 where we saw some of the useful theorems such as linearity, proportionality, superposition, Thévenin's and Norton's equivalents, and matching. All of these still apply as we deal with $Z(s)$ and $H(s)$. I'll use them in examples.

6.1 WORKING WITH Z(s)

Instead of using lots of words to describe what we do in the frequency domain, I'll just do some examples. Then when we finish those, I'll make some general remarks about $Z(s)$. Keep in mind, though, that we are always dealing with systems in the *zero state* so that all initial conditions are zero.

6.1.1 Example I

Fig. 6.1 is a simple circuit in the time domain. I'm going to use this circuit to answer a number of different questions in the s domain.

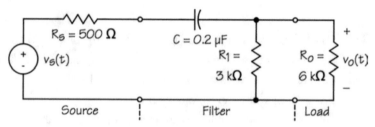

FIGURE 6.1: Example I in the time domain.

The first step in our analysis in the frequency domain is to get the circuit converted. But as I said at the beginning, we are working with *zero-state* circuits, so the conversions from t to s are the simple ones without the extra sources. Resistors R stay

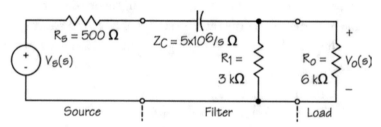

FIGURE 6.2: Example I in the frequency domain.

R, inductors L become Ls, and capacitors C become $1/Cs$. Fig. 6.2 is the transformed circuit.

Question 1: What's the input impedance to the loaded filter? In other words, what is the impedance looking from the source to the right into the filter, including the load R

Treat the elements just as you would handle resistances. Ignore the source. Then the input impedance Z_{in} is Z_C in series with the parallel combination of R_1 and R_o:

$$Z_{in}(s) = Z_C + R_1 \| R_o$$

$$= \frac{\times 10^6}{s} + \frac{3000 \times 6000}{3000 + 6000}$$

$$= 2000 \frac{s + 2\ 00}{s} \, \Omega$$

You are supposed to ask what this means! If you happen to, we can see a few things just by looking at the function $Z(s)$:

- Its unit is ohms, just like resistance.
- It is the ratio of two polynomials in s.
- As s, which is frequency, gets large, the fraction gets closer and closer to 1. So $Z(s)$ starts to look like 2000 Ω. We can see this from the circuit: as the frequency goes up, the impedance of the capacitor decreases ($Z_C = 1/Cs$). Hence the capacitor looks more like a short circuit, leaving only R_1 in parallel with R_o, which is 2 kΩ.

Question 2: What is the impedance looking in from the load with the sources dead? This you recognize is the Thévenin equivalent impedance.

Again, treat this as you would a DC circuit made of just resistors. With the source dead, which means zero voltage, which means a short circuit, the Z_{Th} is the series combination of R_s and Z_C, paralleled by R_1:

$$Z_{Th}(s) = \frac{\left(00 + \dfrac{\times 10^6}{s}\right) 3000}{00 + \dfrac{\times 10_6}{s} + 3000}$$

$$= 42 \quad 6 \frac{s + 10000}{s + 142 \quad 6} \Omega$$

Note again:

- $Z_{Th}(s)$ is a ratio of two polynomials in s.
- As s gets very large, $Z_{Th}(s)$ approaches 428.6 Ω, which is 500 || 3000 (the capacitor again begins to look like a short circuit).
- As s gets very small, which is DC, the capacitor looks like an open circuit. Then $Z_{Th}(s)$ reduces to 3000 kΩ.

Question 3: What's the Thévenin equivalent? We've already done half the work. Let's find the open-circuit voltage via a voltage divider with the load R_o removed:

$$V_{oc}(s) = \frac{3000}{3000 + \dfrac{\times 10_6}{s} + \quad 00} V_s(s)$$

$$= 0 \quad \frac{s V_s(s)}{s + 142 \quad 6}$$

I've left off the units because we have no units for $V_s(s)$. Once we know that, $V_{oc}(s)$ will have the same units. Note also:

- When s gets very large, the output is $0.857\,V_s(s)$, which is what the two resistors provide when the capacitor becomes a short circuit.
- When s goes to zero (DC), V_{oc} also goes to zero, because the capacitor becomes an open circuit.

Question 4: Using the Thévenin equivalent, what is $V_o(s)$? Fig. 6.3 shows the circuit.

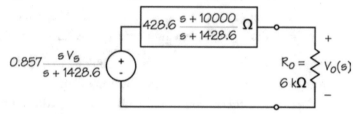

FIGURE 6.3: Thévenin equivalent of Example I.

We can write the output voltage by inspection, just as we did for resistive circuits, using the voltage divider:

$$V_o(s) = 0 \quad \dfrac{sV_s(s)}{s+142\ 6} \dfrac{6000}{6000+42\ 6\dfrac{s+10000}{s+142\ 6}}$$

$$= 0 \quad \dfrac{sV_s(s)}{s+2000}$$

We'll be back to this example later.

6.1.2 Example II

Here's a more complicated circuit that will illustrate more of our work in the s domain. Fig. 6.4 shows the original time-domain circuit.

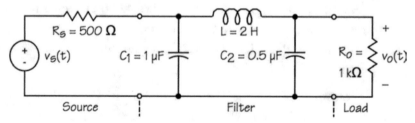

FIGURE 6.4: Example II in the time domain.

Conversion to the frequency domain is easy because we are requiring our circuits to be *zero state*. Fig. 6.5 shows the conversion. I've labeled the circuit for nodal analysis.

To answer questions about this circuit, I'm going to do a complete nodal analysis:

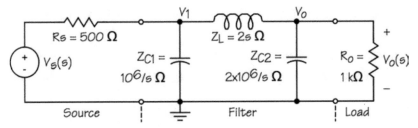

FIGURE 6.5: Example II in the frequency domain.

$$\frac{V_1 - V_s}{R_s} + \frac{V_1}{Z_{C1}} + \frac{V_1 - V_o}{Z_L} = 0,$$

$$\frac{V_o - V_1}{Z_L} + \frac{V_o}{Z_{C2}} + \frac{V_o}{R_o} = 0$$

Solving this with the numbers given in the circuit yields

$$V_o = 2 \times 10 \frac{V_s}{s^3 + 4000s^2 + \quad \times 10^6 s + 3 \times 10},$$

$$V_1 = 2000 \frac{(s^2 + 2000s + 10^6)V_s}{s^3 + 4000s^2 + \quad \times 10^6 s + 3 \times 10}$$

The only thing I really want to note about those right now is that their denominators are the same. Let's pose and answer some questions.

Question 1: How much load does this filter (with its load) put on the source? In other words, we want the input impedance as shown in Fig. 6.6.

In the drawing, the two terminal pairs are designated as *ports*. A port is a pair of terminals that we designate for defining a way into or out of the circuit. So Question 1 asks for the value of the impedance "looking into" Port 1 with the load R_o attached to Port 2.

I have already solved for V_1. I need the current flowing into the port. That current is $(V_s - V_1) / R_s$:

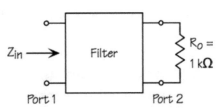

FIGURE 6.6: Example II as block with ports.

$$Z_{in}(s) = \frac{V_1}{I_{in}} = \frac{V_1}{(V_s - V_1)/R_s}$$

$$= 10^6 \frac{s^2 + 2000s + 10^6}{s^3 + 2000s^2 + 1 \quad \times 10^6 s + 10} \Omega$$

Notice that this impedance, like the previous ones, has units of ohms and is the ratio of two polynomials in s.

Question 2: What's the Thévenin equivalent of this circuit as seen from R_o?

I'll start by solving the equations with the R_o term omitted. The result for V_{oc} is

$$V_{oc} = \frac{2 \times 10 \; V_s}{s^3 + 2000 s^2 + 1 \; \times 10^6 s + 2 \times 10}$$

I find Z_{Th} by replacing the voltage source with a short circuit and using parallel and series elements to give

$$Z_{Th} = (R_s \| Z_{C1} + Z_L) \| Z_{C2}$$
$$= \frac{2 \times 10^6 (s^2 + 2000 s + 0.5 \times 10^6)}{s^3 + 2000 s^2 + 1.5 \times 10^6 s + 2 \times 10^9} \; \Omega.$$

Question 3: Can you build this Thévenin equivalent using real Rs, Ls, and Cs? Sure, if you remember that this all came from the circuit in Fig. 6.4! After all, that's what this equivalent is equivalent to.

But suppose you don't know that? Suppose the two equations for the Thévenin equivalent simply appeared on your front porch begging to have a real circuit drawn for them? Can you find such a circuit?

Maybe! There is no guarantee that a particular impedance can be reduced to actual circuit elements. We'll see more of this question later, though.

6.1.3 Some Things to Note About $Z(s)$

Impedances are functions $Z(s)$, where s is the frequency in radians per second. But $Z(s)$ was developed from real circuit elements, or at least from ideal models. All of these models are linear. So there are some things that we can notice by just looking at $Z(s)$ and thinking a little.

First, R, L, and C are always positive numbers. We don't have negative resistors, at least without having nonlinear elements of some kind. We don't have negative inductors or capacitors, either. Therefore no combination of these, be it series, parallel, or whatever, can yield a minus sign anywhere. So $Z(s)$ can't have minus signs in it.

You have already probably noticed that the $Z(s)$ results that we have gotten in our examples seem to come out to be ratios of polynomials in s. In fact, that's what's always going to happen.

The practical reason is that we are combining our circuit elements after transforming them into the frequency domain. So our elements have impedances of R, Ls, and $1/Cs$. The result is polynomials in *s*. And since one-over-*s* appears in element impedances, we should expect to have *s* in both the numerator and the denominator of Z(*s*).

Now we can conclude, based on our observations so far, that Z(*s*) is the ratio of two polynomials in *s* and that there are no minus signs in either of these polynomials. (Well, technically, *every* term could be negative in both the numerator and the denominator, but then we'd "cancel" these minus signs and every term would then be plus.)

The denominator of Z(*s*) is important because, once you get it, it's hard to get rid of. No matter where we measure an impedance in a circuit, the resultant Z(*s*) will have the same denominator polynomial. This polynomial is *characteristic* of the circuit itself and is so named. I'll leave to other courses the fact that we can determine the stability of a system by looking at its characteristic polynomial.

An extension of our thinking says that we can write functions Z(*s*) which cannot be created using passive elements. For example, a Z(*s*) that we think up that has a minus sign in it cannot be modeled using linear resistors, capacitors, and inductors.

Z(*s*) is a function of *s*, so it's tempting to think of this as being somehow related to the Laplace transform. After all, we did derive the *s*-domain representations of the circuit elements through the Laplace transform. But Z(*s*) is *not* the Laplace transform of anything.

Consider the following impedance, which seems to have the characteristics we've already noted:

$$Z(s) = 24 \frac{s^2 + 300s + 12\ 00}{s^3 +\ 00s^2 +\ 0000s + 10^6} \cdot \Omega$$

What happens if I don't understand this bit about the inverse transform and do it anyhow? The result is a function in the time domain:

$$z(t) = e^{-200t} (33 \quad 100t + \quad 100t) - \quad e^{-100t}\ \Omega$$

While this might have been a fun exercise in inverse Laplace transforms, the result as an impedance is 100% meaningless! Is this really an impedance that is a function of time? Not if we have built the circuit out of ordinary, fixed resistors, capacitors, and inductors.

Impedance is a *frequency-domain* concept, and it's for circuits in the zero state at that.

6.1.4   One More: Example III

Let's do one more impedance example by analyzing the circuit of Fig. 6.7.
Fig. 6.8 shows this circuit converted to the frequency domain.

What is the impedance looking into the circuit from the left when a 1-kΩ load is connected at the right? Fig. 6.9 shows the connections. I have labeled the circuit for nodal analysis while preserving the definition of V_2 for later use. Note, however, that V_1 and V_2 do not have a common–terminal; I'll need a node voltage at the bottom right, which I'm calling V_x. (See Fig. 6.9.) So that I don't lose sight of V_2, I'll keep it and then label the top right node relative to V_x.

Node equations can be written at the two nodes on the right:

$$\frac{(V_x + V_2) - V_s}{2000} + \frac{V_x + V_2}{10^6/s} + \frac{V_2}{1000} = 0,$$

$$\frac{V_x}{00} + \frac{V_x - V_s}{10^6/s} - \frac{V_2}{1000} = 0$$

Since I want the input impedance, which is the impedance seen by the source V_s, I need the current flowing to the right at the V_s node. That current, in terms of node voltages, is

$$I_s = \frac{V_s - (V_x + V_2)}{2000} + \frac{V_s - V_x}{10^6/s}$$

I'll solve these equations for I_s and V_2 and then compute Z_{in} as V_s / I_s, yielding

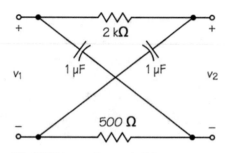

FIGURE 6.7: Example III in the time domain.

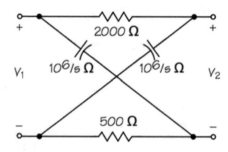

FIGURE 6.8: Example III in the frequency domain.

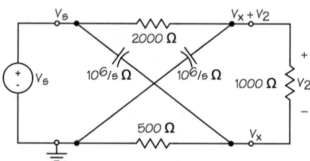

FIGURE 6.9: Example III loaded.

$$I_s = \frac{V_s}{2000} \frac{s + 2000}{s + 3\ 00},$$

$$V_2 = \frac{-V_s(s - 1000)}{s + 3\ 00},$$

$$Z_{in} = 2000 \frac{s + 3\ 00}{s + 2000} \Omega$$

Does Z_{in} have the characteristics we've already noted? It's a ratio of polynomials in s with no minus signs and it has the unit of ohms. That all looks right.

But how about V_2? There are minus signs! But is this wrong? No, because V_2 is not an impedance. Rather, it is a voltage that depends on the input from the source V_s. If we knew V_s, we could compute V_2 and then use the inverse Laplace transform to return V_2 to the time domain as $v_2(t)$. But even in the expression for V_2 there are no minus signs in the denominator.

6.2   TRANSFER FUNCTIONS $H(s)$

So far we've been looking at impedances $Z(s)$, which are ratios of voltage to current in a circuit, expressed in the frequency or s-domain. Our circuits have been in the zero state because the initial conditions are all zero.

Impedances aren't the only functions we work with when we analyze circuits in the frequency domain. We often want to know what happens at one place in the circuit as a result of what we do at another place. There are two such relationships that we could investigate, transfer impedances and transfer functions.

Transfer impedance is the ratio of the voltage across some circuit element to the current applied somewhere else. In the example in the previous section, we could find the voltage V_2 as a result of the current I_s and call this the transfer impedance $Z_{2s}(s)$:

$$Z_{2s}(s) = \frac{V_2}{I_s} = \frac{-V_s(s-1000)}{s+3\ 00} \frac{2000(s+3\ 00)}{V_s(\ s+2000)}$$

$$= \frac{2000(s-1000)}{s+2000} \Omega$$

Note that *transfer* impedances probably don't have the sign characteristic that we saw for... hmmm, what do we call the kind of impedances we've just been studying? For those impedances, we observed the voltage and the current at the *same* point, so we call these *driving-point* impedances.

One other characteristic is noticeable, the *characteristic polynomial*. The denominator of Z_{2s} is the same as the denominator of Z_{in}.

The other relationship that we often want to use is the *transfer function*. This function, still in the s-domain and still for zero-state circuits, is the ratio of the voltage at some point to the voltage at another point. These two points are generally the output and the input ports of a circuit. (The transfer function could also be the ratio of two currents.)

Let's study transfer functions by working some examples.

6.2.1 Example IV

Go back to Example I, shown here in Fig. 6.10. This time I'd like to find the output voltage $V_o(s)$ as a function of the input voltage $V_s(s)$. For variety, I'll use proportionality, so I've marked the circuit with some voltages and currents.

I'll start by assuming that $V_o = 1$ and track that through to the input:

$$V_o = 1,$$
$$I_o = 1/6000,$$
$$I_1 = 1/3000,$$
$$I_C = 1/6000 + 1/3000 = 1/2000,$$
$$V_C = \frac{\times 10^6}{s} I_C = 2\ 00/s,$$
$$V_R = \ 00 I_C = 0\ 2\ ,$$
$$V_s = V_o + V_C + V_R = 1\ 2\ + 2\ 00/s,$$
$$\frac{V_o}{V_s} = \frac{1}{1\ 2\ + 2\ 00/s} = 0\ \quad \frac{s}{s+2000}$$

This output:input voltage ratio is often labeled $H(s)$. This function is important in a number of areas of study of electrical systems. Why? I'll make a simple-minded statement: If $V_s(s) = 1$, then the output $V_o(s) = H(s)$.

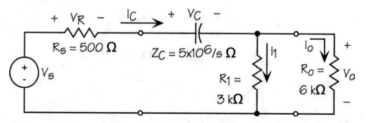

FIGURE 6.10: Example I revisited as Example IV.

Right! So what? Well, what is the input $v_s(t)$ if $V_s(s) = 1$? Indeed, the input must be the unit impulse $v_s(t) = \delta(t)$ because the Laplace transform of the unit impulse is 1. But what good is this? After all, there isn't such a thing as a real impulse.

But the unit impulse is the simplest of the sources we can have when it is transformed into the s-domain (i.e., just 1). So we use the "impulse response" as a touchstone measure of the response of a system.

The impulse response is $h(t)$, a time-domain function. Its transform is the transfer function $H(s)$. For the circuit of Example IV, the transfer function is

$$H(s) = 0 \quad \frac{s}{s+2000}$$

Now let's do some things with $H(s)$ by trying a few inputs and seeing both the frequency-domain and time-domain outputs.

Input 1. Start with the impulse itself by making the input the unit impulse:

$$v_s(t) = \delta(t),$$
$$V_s(s) = 1,$$
$$V_o(s) = H(s)V_s(s) = 0 \quad \frac{s}{s+2000},$$
$$v_o(t) = 0 \quad \delta(t) - 1600e^{-2000t}u(t)$$

So the output, which is the impulse response, itself contains an impulse and then a term that dies out. Can we see what's happening in the circuit itself? Sure! At $t = 0^-$ (just before the impulse blasts off), the capacitor is uncharged. So its voltage at $t = 0^-$ is 0. Hence the impulse itself appears in the output as a result of a simple voltage divider.

Where does the -1600 term come from? The impulse also makes an infinite current flow (for zero time!), with the result that the capacitor is charged to 2000V. How could that be? Look at the math. At $t = 0^-$, the capacitor is uncharged. When the impulse hits, the current in the circuit $i_C(t)$ will be

$$i_C(t) = \frac{v_s(t)}{R_s + R_1 \| R_o} = \frac{\delta(t)}{2\ 00}$$

We can integrate the current to get the capacitor voltage:

$$v_C(t) = \frac{1}{C} \int_{0^-}^{t} i_C(x)dx = \frac{1}{0\ 2\times10^{-6}} \int_{0^-}^{t} \frac{\delta(x)}{2\ 00} dx$$

The integrand has a value only at $t = 0$, so this integration can be split into two parts:

$$v_C(t) = \frac{\times10^6}{2\ 00} \left(\int_{0^-}^{0^+} \delta(x)dx + \int_{0+}^{t} \delta(x)dx \right)$$

The first integration will produce 1 because that's the area under the unit impulse. The second integration will produce 0 because the impulse has already passed. So the result is

$$v_C(t) = 2000(1+0) = 2000$$

After $t = 0$ this capacitor voltage drives a current in the counterclockwise direction through R_s and the parallel combination of R_1 and R_o. The result is that $v_o(t)$ starts at -1600 V and decays to 0.

Gosh, it's taken half a page to show that the frequency-domain approach to finding the time-domain solution gives a sensible result!

Input 2. Let's see what happens when we apply a step of 10 V:

$$v_s(t) = 10u(t) \quad,$$
$$V_s(s) = 10/s,$$
$$V_o(s) = H(s)V_s(s) = \frac{}{s + 2000},$$
$$v_o(t) = e^{-2000t}u(t)$$

No surprises there. Of course, we can't really have perfect steps in the physical world, but we can come as close as we please (or have money for).

One matter about units: I've ignored the units on $V_s(s)$. The "10" carries the unit of volts. s has the unit of "per second." So the unit of $V_s(s)$ is "volt-seconds." But we get sloppy and simply say "volts."

Input 3. Now make the input a sinusoid starting at $t = 0$:

$$v_s(t) = 20\cos(2000t)u(t)\,\text{V},$$
$$V_s(s) = \frac{20s}{s^2 + 2000^2},$$
$$V_o(s) = 0.8\frac{s}{s + 2000}\frac{20s}{s^2 + 2000^2},$$
$$= \frac{16s^2}{(s + 2000)(s^2 + 2000^2)},$$
$$v_o(t) = \left(8e^{-2000t} + 8\cos 2000t - 8\sin 2000t\right)u(t)\,\text{V}$$
$$= \left[8e^{-2000t} + 8\sqrt{2}\cos(2000t + 45°)\right]u(t)\,\text{V}.$$

Note that this response includes two parts. The first term dies out, while the second goes on forever. The plot in Fig. 6.11 shows these two pieces and the total response.

Keep in mind that we can recognize these two parts in the frequency-domain response. The part that dies out will have factors in the denominator that have real roots or roots that have real parts. The parts that don't die out have roots that are purely imaginary.

A partial-fraction expansion of $V_o(s)$ shows these clearly:

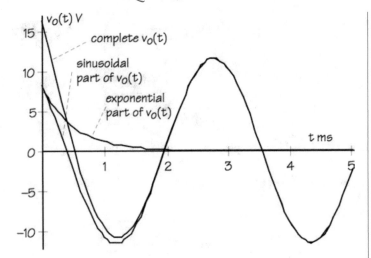

FIGURE 6.11: Output of Example IV.

$$V_o(s) = \frac{16s^2}{(s+2000)(s^2+2000^2)}$$

$$= \frac{8}{s+2000} + \frac{8s}{s^2+2000^2} - \frac{8 \times 2000}{s^2+2000^2}.$$

The first term transforms to a decaying exponential. The second and third terms are a cosine and a sine respectively; they don't die out.

You should recognize the "sinusoidal part of $v_o(t)$" as the sinusoidal steady state. We will spend more time with this in the next chapter.

6.2.2 Example V

Now let's revisit Example II, shown in Fig. 6.12, to find its impulse response. We've already solved this problem using nodal analysis:

$$V_o = 2 \times 10 \frac{V_s}{s^3 + 4000s^2 + \quad \times 10^6 s + 3 \times 10},$$

$$V_1 = 2000 \frac{(s^2 + 2000s + 10^6)V_s}{s^3 + 4000s^2 + \quad \times 10^6 s + 3 \times 10}$$

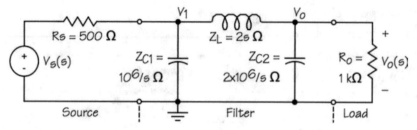

FIGURE 6.12: Example II revisited as Example V.

Getting $H(s)$ is easy:

$$H(s) = \frac{V_o}{V_s} = \frac{2\times10}{s^3 + 4000s^2 + \quad \times10^6 s + 3\times10}$$

And that's a good stopping place.

6.2.3 Some Things to Note About $H(s)$

$H(s)$ is a transfer ratio, generally the ratio o f output voltage to input voltage. When the input is 1 in the s-domain, the output equals $H(s)$.

Within the mathematical function for the output $V_o(s)$ we can always separate the part that is due to the input from the part that is due to the natural response of the circuit. We will do this using partial fractions when we look at the sinusoidal steady state in the next chapter.

You have noticed, I'm sure, that $H(s)$ is a ratio of polynomials in s, just as $Z(s)$ is. One significant difference, though, is that minus signs can appear in the numerator. Recall the output voltage for Example III (the circuit that looked like an X):

$$V_2(s) = \frac{-V_s(s-1000)}{s+3\ 00},$$

$$H(s) = \frac{-(s-1000)}{s+3\ 00}$$

That's got some minus signs, but *none in the denominator*. As we shall see later, all the roots of the denominator polynomial must have negative real parts, which means the coefficients of that polynomial in s must *all* be positive. (Well, yes, they could *all* be negative.)

6.3   THE *S*-PLANE

The *s*-plane is an important concept in a number of areas of electrical engineering. It's a graphical device for "seeing" $Z(s)$ and $H(s)$. It gives us a way of estimating the overall response of a system. Through it we can answer questions about stability. We can determine how the system behaves when excited by sinusoids, which we will do in Chapter 7.

The *s*-plane can also be used for graphical constructions to get numerical answers. In the pre-computer age we often did this; today it isn't of much use. We'll do a couple of examples, though, to help get across the ideas of what the *s*-plane represents.

Most of what we will do in this section has the goal of helping us visualize how a circuit behaves so that we get a better understanding of its operation.

6.3.1   What is *S*?

The independent variable of the frequency domain is *s*. Recall that *s*, as in $H(s)$, is the frequency and has the units of "per second." *s* is a complex number and hence has both real and imaginary parts.

The real part of *s* is sigma, σ, with units of s^{-1}. Since some people are uncomfortable saying "12.5 per second," the unit is sometimes said to be "nepers per second." "Neper" is a unitless quantity and is named after John Napier, who invented logarithms. (There is some question about how his name is properly spelled!) Sigma is sometimes called the "neper frequency," but that's fairly rare today.

The imaginary part of *s* is omega, ω, with units of s^{-1}. More commonly, though, we say "radians per second." "Radian" is unitless also.[1] Omega is often called the "radian frequency."

Consider e^{st} that arises in just about everything we do. Let's replace *s* with $\sigma + j\omega$ and manipulate things a little using Euler's formula:

$$e^{st} = e^{(\sigma + j\omega)t} = e^{\sigma t}e^{j\omega t}$$
$$= e^{\sigma t}(\quad \omega t + j \quad \omega t)$$

---

[1] The *radian* is said to be named after Arthur Radian. While inventing tires, he developed a method of measuring a tire's diameter. He made a wedge-shaped template that subtended an angle of exactly 57.29577951°. This he placed on the tire with its point exactly at the tire's center. He then laid a flexible tape measure on the circumference, measured the distance between the two edges of his wedge, and doubled the result to obtain the tire's diameter. It is believed by some that the radial tire is also named in Arthur's honor. Radian claimed to have been born in Center Point, Indiana.

Note that σ provides damping, which means reducing the size of the sinusoidal term as time goes on. This term *must* decay in any bounded real system, so we will always see that $\sigma \leq 0$ if our circuits are to be made with real components.

Note also that $j\omega$ leads to the sinusoidal terms that don't die out, at least not by themselves.

6.3.2 What is the S-Plane?

Very simply, the *s*-plane is the plane defined by the σ and the $j\omega$ axes. But let's do something else first. Go back to Example I, where $Z_{in}(s)$ was the ratio of two polynomials in *s*:

$$Z_{in}(s) = \frac{2000(s + 2\ 00)}{s} \Omega$$

The numerator has the term $s + 2500$, so the root of the numerator polynomial is $s = -2500$. We can write this as $s = -2500 + j0$. When *s* takes on this value, the numerator becomes zero, so $Z_{in}(-2500) = 0$. We say that this is a *singularity* of $Z_{in}(s)$ because it is a place where $Z_{in}(s)$ goes to zero. We call $s = -2500 + j0$ a *zero of* $Z_{in}(s)$.

Similarly, the denominator has the root $s = 0$, which is $s = 0 + j0$. When *s* takes on this value, the denominator becomes zero, so $Z_{in}(0) = \infty$. This is also a *singularity* of $Z_{in}(s)$ because it is a place where $Z_{in}(s)$ becomes infinite. We call $s = 0 + j0$ a *pole* of $Z_{in}(s)$.

Let's plot $Z_{in}(s)$ in three dimensions in Fig. 6.13.

Take a close look at the drawing and note several things:

- The pole at $s = 0 + j0$ is pretty obvious! I plotted this on a scale that chopped off the top of the pole (an infinite amount of chop, for that matter). In Fig. 6.13 this looks like an artistic chimney.
- The zero at $s = -2500 + j0$ is a little harder to see—it's that rather broad depression to the northwest of the pole.
- The general level of the surface far away from the singularities is

$$Z_{in}(s) = \frac{2000(s + 2\ 00)}{s} = 2000 \ \Omega$$
$$\scriptsize s \to \infty \qquad\qquad s \to \infty$$

The dotted line (hard to see) along the left front face indicates this general level of 2 kΩ.

Three-dimensional plots are messy to generate. Besides, it's not the general surface that tells the story, it's the singularities, the poles and zeros. These tell us something important about the overall response of the circuit.

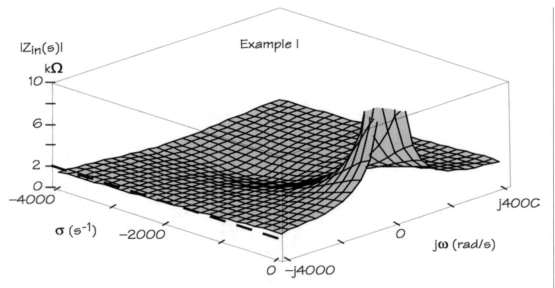

FIGURE 6.13: 3D ploat of $Z_{in}(s)$.

So how about 2D? Let's draw just the plane that underlies the whole 3D plot. Let's draw the plane that is at $|Z_{in}(s)| = 0$. Fig. 6.14 is that plane.

In Fig. 6.14, the circle indicates a *zero* of $Z_{in}(s)$; this is at $s = -2500 + j0$. The cross indicates a *pole* of $Z_{in}(s)$; this is at the origin.

We can get both the magnitude and the angle of $Z_{in}(s)$ for any s from this s-plane— almost. Suppose we want the magnitude and the angle of $Z_{in}(s)$ for $s_1 = -1000 + j2000$ s^{-1}. The graphical construction for this is shown in Fig. 6.15.

I have drawn lines (vectors?) from the singularities to the point s_1 where $s = -1000 + j2000$ s^{-1}. I will find the magnitude of Z_{in} at this frequency by using the lengths of these lines:

$$s) = \frac{2000(s+2\ 00)}{s} \Big|_{s \to \infty} = 2000\ \Omega$$

$$\left|Z_{in}(\ 1000 + j2000)\right| = \frac{|B|}{|A|}$$

$$= \frac{\sqrt{(-1000-(-2\ 00))^2 + 2000^2}}{\sqrt{1000^2 + 2000^2}}$$

$$= \frac{2\ 00}{2236\ 1} = 1\ 11\ \Omega$$

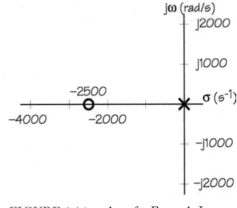

FIGURE 6.14: s-plane for Example I.

Ooops! The s-plane doesn't carry the scale factor for the surface above the s-plane. Here we have already found that it is 2000, so

$$\left| Z_{in}(-1000 + j2000) \right| = 2000 \times 1.11$$
$$= 2236 \ \Omega$$

I can find the angle of Z_{in} the same way:

$$\angle Z_{in}(-1000 + j2000) = \angle B - \angle A$$

$$= \tan^{-1} \frac{2000}{-1000-(-2\ 00)} - \tan^{-1} \frac{2000}{-1000}$$

$$= 1\ 1° - 116\ 6° = -63\quad °$$

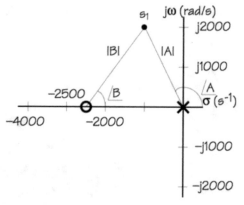

FIGURE 6.15: Construction of $Z_{in}(-1000 + j20000)$.

So $Z_{in}(-1000+j2000) = 2236\underline{/-63.5°} \ \Omega$. But what's the point? My point is that the s-plane carries *all* the information about $Z_{in}(s)$ except the magnitude scale factor. This is a constant that must be determined just once. Hence the 3D plot is not needed once you understand what the s-plane shows.

6.3.3 Another Z(s) Example

Example II earlier in this chapter has an input impedance of

$$Z_{in}(s) = \frac{10^6(s^2 + 2000s + 10^6)}{s^3 + 2000s^2 + 1\ \times 10^6 s + 10} \ \Omega$$

Factor this to show all the roots:

$$Z_{in}(s) = \frac{10^6(s + 1000)^2}{(s+1441)(s+2\ 0 + j\quad)(s+2\ 0 - j\quad)} \ \Omega$$

Now we can see where all the singularities are:

- double zero (the numerator factor is squared) at $s = -1000$ s^{-1};
- poles at $s = -1441$ and $-280 \pm j785$ s^{-1}.

For very large s, $Z_{in}(s)$ goes to zero.

    Fig. 6.16 is the 3D plot of this Z_{in}. The s-plane for this is shown in Fig. 6.17.

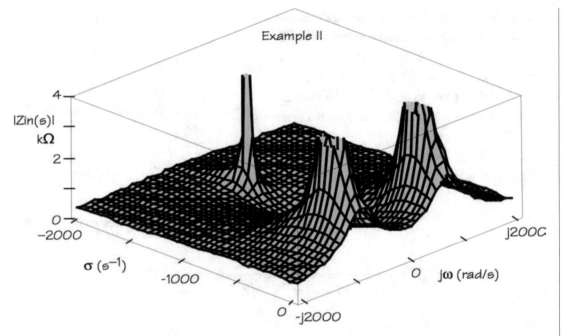

FIGURE 6.16: 3D plot for Example II.

Let's use this *s*-plane to estimate where $Z_{in}(s)$ is largest on the $j\omega$ axis. (We conventionally stick to the positive $j\omega$ axis. The negative half mirrors the positive half because complex poles always occur in conjugate pairs.)

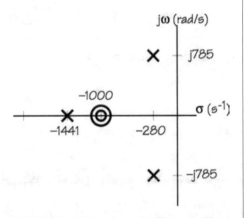

FIGURE 6.17: *s*-plane for Example II.

An inspection of the *s*-plane should lead us to the conclusion that the largest $Z_{in}(j\omega)$ will be on the $j\omega$ axis close to the pole at $j785$ rad/s. After all, the pole represents a value of *s* that makes this impedance infinite.

I'm going to guess that the maximum is at $j800$ rad/s. Fig. 6.18 shows vectors from each singularity to $j800$ rad/s. The line from the double zero is really two lines.

Now calculate $|Z_{in}(j800)|$:

$$|Z_{in}(j800)| = \frac{\left(\sqrt{1000^2 + 800^2}\right)^2 K}{\sqrt{1441^2 + 800^2}\sqrt{(800 - 785)^2 + 280^2}\sqrt{(800 + 785)^2 + 280^2}}$$

$$= 2.205 \times 10^{-3} K.$$

Remember that we don't know the scale factor K, but I can get it by a simple calculation. I'll evaluate $|Z_{in}(0)|$ both from the s-plane and from the function itself. I choose $s = 0$ because that's a simple place to do the job. (I'd have to choose somewhere else if there had been a zero at $s = 0$.)

From the s-plane (drawing lines from each singularity to the origin):

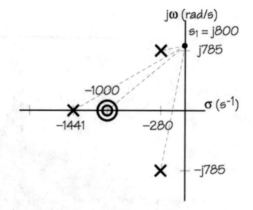

FIGURE 6.18: s-plane construction for $j800$ rad/s.

$$|Z_{in}(0)| = \frac{1000^2 K}{(1441)\sqrt{2\ 0^2 +\quad}^2 \sqrt{2\ 0^2 +\quad}^2} = 10^{-3} K$$

From the functional form of Z_{in} I get

$$|Z_{in}(0)| = \frac{10^6(s^2 + 2000s + 10^6)}{s^3 + 2000s^2 + 1\ \times 10^6 s + 10}\bigg|_{s=0} = 10^3\ \Omega$$

Now equate these to find K:

$$10^{-3} K = 10^3\ \Omega,\ so\ K = 10^6\ \Omega$$

Using this,

$$|Z_{in}(j\ 00)| = 2\ 20\ \times 10^{-3} \times 10^6 = 220\ \Omega$$

(The actual peak of 2209 Ω is at $\omega = 790$ rad/s.)

Here again, my point is not that one can do magnificent calculations graphically, but rather that the s-plane can be used to get a quick estimate of the behavior of a function plotted on it.

6.3.4 An Example of $H(s)$

How about applying our s-plane knowledge to a transfer function $H(s)$. Remember that $H(s)$ is the ratio of the output to the input. Example III, developed earlier in this chapter, is a good one to work with.

From our earlier work, we got

$$V_2 = \frac{-V_s(s-1000)}{s+3\ 00},$$

which leads to

$$H(s) = \frac{V_2}{V_s} = \frac{-(s-1000)}{s+3\ 00}$$

Fig. 6.19 is $H(s)$ in 3D. The zero is hard to see; it is a shallow dip on the *right* side of the $j\omega$ axis.

When we plot the *s*-plane (Fig. 6.20), we see both of the singularities clearly.

The zero is in the right half-plane, i.e., it has a positive real part. It's OK to have a zero there, but we'll see in the next example that it isn't OK to have a pole there.

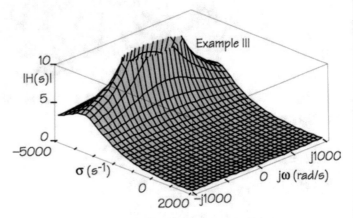

FIGURE 6.19: 3D plot of $H(s)$ for Example III.

6.3.5   Another $H(s)$ Example

Consider the following $H(s)$:

$$H(s) = 1000\frac{s+200}{s^2-200s+\ 0000}$$

$$= \frac{1000(s+200)}{(s-100+j200)(s-100-j200)}$$

The 3D plot is in Fig. 6.21; there is a zero pretty well hidden by the poles.

Fig. 6.22 shows the *s*-plane for $H(s)$, and the zero is clearly visible.

A pole in the right half-plane of the

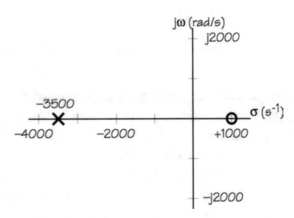

FIGURE 6.20: *s*-plane for Example III.

s-plane says, loud and clear, that the system is unstable! Let's see why this is by going back to the time domain. I'll excite the circuit with a step function, so

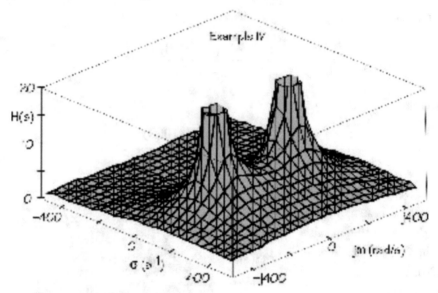

FIGURE 6.21: 3D plot of problematic $H(s)$.

$$v_{in}(t) = u(t),$$

$$V_{in}(s) = 1/s,$$

$$H(s) = \frac{V_{out}(s)}{V_{in}(s)} = 1000\frac{s+200}{s^2 - 200s + 0000},$$

$$V_{out}(s) = 1000\frac{s+200}{s(s^2 - 200s + 0000)},$$

$$v_{out}(t) = \left[4 + e^{+100t}(\quad 200t - 4 \quad 200t)\right]u(t)$$

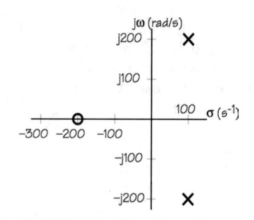

FIGURE 6.22: s-plane for problematic $H(s)$.

What did I do? Given $v_{in}(t)$, I transformed this to the frequency domain and then multiplied $H(s)$ by it to get the output $V_{out}(s)$. The inverse Laplace transform took me back to the time domain.

The problem with this $H(s)$ is very evident in the time-domain result. Notice the term e^{+100t}. This says that the step input creates an output that grows without bound! In other words, the output is going to be infinite.

So we can't have a stable system if there are any poles in the right half-plane (the plane to the right of the $j\omega$ axis) of the transfer function $H(s)$. In fact, the system can't be completely stable if there are any poles *on* the $j\omega$-axis, either. (We often abbreviate right half-plane as RHP, and this is a good time to ask about right-half-plane joke.)

6.3.6   What's This All About?

The s-plane is a very useful way of representing information about a system. In particular,

- the s-plane is neat for visualizing how a system will respond to various inputs;
- the singularities shown on the s-plane exert considerable control over that response;
- the poles of a function govern the time-domain response because they form the individual denominators in a partial-fraction decomposition; and
- poles are not in the right half-plane (RHP) if the system is going to be stable.

6.4   EXTRA EXAMPLES

These examples all deal with functions of s. They include plotting s-planes, finding a Thévenin equivalent, and designing a circuit to fit a particular s-plane pole-zero constellation.

6.4.1   Example VI

For the circuit of Fig. 6.23, find $Z(s)$ and sketch the pole-zero constellation on the s-plane.

The first step is to find the impedances of the inductor and the capacitor:

$$Z_L = 0 \times 10^{-3} s\ \Omega,$$
$$Z_C = 1/0\ 022 \times 10^{-6} s\ \Omega$$

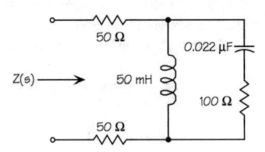

FIGURE 6.23: Example VI: circuit.

Now I can find $Z(s)$ by finding Z_L in parallel with the Z_C-R combination, then adding in series the two 50-Ω resistors. The result, given in the proper form of the ratio of two polynomials in s, is

$$Z(s) = 0 + Z_L \| (Z_C + 100) + 0$$

$$= 200 \frac{s^2 + 22\ 2\ 0s + 0\ 4\ 4\ \times 10}{s^2 + 2000s + 0\ 0\ 10 \times 10}\ \Omega$$

Note that the highest power of s in both polynomials has a coefficient of 1.

The zeros of $Z(s)$ are the roots of the numerator polynomial; the poles, of the denominator:

$$Z_{zeros} = -200 \,, -2262\ 0 \quad^{-1},$$

$$Z_{poles} = -1000 \pm j3013 \quad^{-1}$$

$Z(s)$ has two zeros on the negative real axis and a pair of complex conjugate poles. Don't forget that complex poles always must occur in conjugate pairs. The s-plane is shown (not to scale) in Fig. 6.24. Note that both axes are labeled with units.

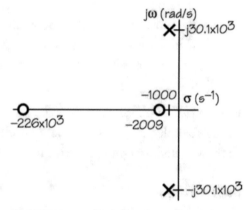

FIGURE 6.24: Example IV: s-plane.

6.4.2 Example VII

The circuit of Fig. 6.25 is to be considered a source that can have a Thévenin equivalent.

For a Thévenin equivalent, I need any two of the following: open-circuit voltage, short-circuit current, and Thévenin equivalent impedance. The last one is found by "looking into" the circuit with all independent sources dead.

Let's start by finding the impedance of the two energy-storage elements:

$$Z_L = 100 \times 10^{-3} s\ \Omega,$$

$$Z_C = 1/0\ 0 \ \times 10^{-6} s\ \Omega$$

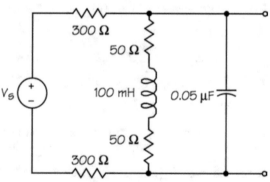

FIGURE 6.25: Example VII: Thévein.

To find the impedance "looking in," I'll first replace the voltage source with a short circuit. Then I can find the impedance between the terminals as the parallel combination of the L branch and the C branch (called Z_o in the equations that follow) and combining that in parallel with the two resistors in series.

The Thévenin equivalent impedance in proper form is

$$Z_o = Z_C \| (\ 0 + Z_L +\ 0),$$

$$Z_{Th} = Z_o \| (300 + 300)$$

$$= 20 \times 10^6 \frac{s + 1000}{s^2 + 34333s + 233\ 3 \times 10^6}\ \Omega$$

I'll find the open-circuit voltage using a voltage divider. The output of the voltage divider is Z_o that I just computed; the other arm of the divider is the two 300-Ω resistors. The result is

$$V_{oc} = V_s \frac{Z_o}{Z_o + 300 + 300}$$

$$= 33,333 V_s \frac{s + 1000}{s^2 + 34333s + 233\ 3 \times 10^6}$$

That's as far as I can go without knowing something about the source V_s, something like frequency or amplitude. I have V_{oc} and I have Z_{Th} and that's the Thévenin equivalent.

6.4.3   Example VIII

Now find the Thévenin equivalent of Example VII for a particular source voltage and frequency. Then convert the equivalent to the time domain.

The source is to be

$$V_s = 10\,V, f =\ 000\quad,$$
$$s = j2\pi f = j31416$$

Putting these values into the previous results yields the s-domain Thévenin equivalent of

$$V_{oc} = 6\,3\ -j4\ 66 =\ 63\angle{-36}\ ^\circ\quad,$$
$$Z_{Th} = 3\ 2\ -j2\ 6\,0\,\Omega$$

Returning this to the time domain yields

$$v_{oc}(t) =\ 63\quad(2\pi\ 000t - 36\ ^\circ)$$

The Thévenin equivalent impedance can also be returned to the time domain:

$$R_{Th} = 3\ 2\ \Omega,$$

$$\frac{1}{j2\pi\ 000 C_{Th}} = -j2\ 6\,0, C_{Th} = 0\ 111\,\mu$$

Finally in this example, choose a resistive match to the equivalent and find the voltage across that resistor.

The "best" match to a Thévenin impedance is the complex conjugate of that impedance. If the load is restricted to being resistive (i.e., pure real), then the appropriate match is to the magnitude of the equivalent Z_{Th}:

$$R_{match} = |Z_{Th}| = 44 \quad \Omega$$

I can find the voltage across this resistance through a voltage divider consisting of the resistor and the Thévenin equivalent impedance:

$$V_o = V_{oc} \frac{R_{match}}{R_{match} + Z_{Th}} = 3 \quad 1 - j1\ 323$$
$$= 4\ 1 \quad \angle -1\quad 4°$$

This becomes, in the time domain:

$$v_{oc}(t) = 4\ 1 \quad (2\pi\ 000t - 1\quad 4°)$$

6.4.4 Example IX

The circuit of the previous examples is shown in Fig. 6.26 without the source. Our job is to find $H(s)$, the voltage transfer function for the circuit.

In Example VII, I computed the output voltage V_{oc} while I was finding the Thévenin equivalent. That output voltage has V_s as a parameter. So $H(s)$ is just the function V_{oc} from that example divided by V_s.

$H(s)$ and its poles and zeros are

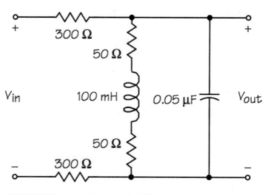

FIGURE 6.26: Example IX: circuit.

$$H(s) = \frac{V_{oc}}{V_s} = 33{,}333 \frac{s + 1000}{s^2 + 34333s + 233\ 3 \times 10^6},$$

$$H_{zero} = -1000 \quad ^{-1},$$

$$H_{poles} = -\ 333, -2\ ,000 \quad ^{-1}$$

Fig. 6.27 shows the s-plane for this $H(s)$.

6.4.5 Example X

For the last example, let's reverse the process. Fig. 6.28 is an s-plane of an impedance. I want a possible circuit.

But the s-plane is always missing a scale factor that tells us the magnitude of the impedance. In this example, we are told that the impedance has a magnitude of 600 Ω at a frequency of 20 krad/s.

Hmmm, a zero at the origin says that the impedance has to look like a short circuit at DC. That means there could be an inductor across the terminals.

Now, let's see…the s-plane has one pole and one zero—equal numbers of poles and zeros. That says there is neither a pole nor a zero at infinite s. Uh, that says the impedance for very large s is a constant value. A resistor in parallel with the inductor could do that job.

I'll start with an impedance as shown in Fig. 6.29.

The impedance of this circuit is

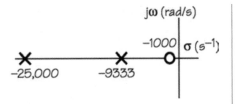

FIGURE 6.27: Example IX: s-plane.

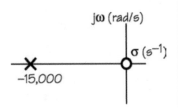

FIGURE 6.28: Example X: s-plane.

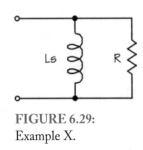

FIGURE 6.29: Example X.

$$Z(s) = Ls \| R = \frac{Rs}{s + \dfrac{R}{L}}$$

So far so good. There's a zero at $s = 0$ and there's a pole at $-R/L$. That means that

$$\frac{R}{L} = 1\ 000, R = 1\ 000L$$

Whee! That gets me R in terms of L. All I have to do is choose one of them so that the magnitude of the impedance is 600 Ω when $s = j20{,}000$ rad/s. I'll write the impedance again, this time putting the newly found value of R into the function and replacing s with $j20{,}000$:

$$\left| Z(j20000) \right| = \left| \frac{1\ 000L(j20000)}{j20000 + 1\ 000} \right|$$

$$= 12000L = 600\ \Omega$$

Solving that for L and then finding R yields

$$L = 0 \quad,$$
$$R = 0\,\Omega$$

The desired impedance consists of a 50-mH inductor and a 750-Ω resistor in parallel. (Why anyone would want to do this is a mystery, but it does provide a very straightforward design example!)

6.5   CIRCUIT DESIGN EXAMPLE

Design a $Z(s)$ that will have the poles and zeros shown on the s-plane of Fig. 6.30. The zero may be anywhere on the sigma axis except at the origin. At $s = 0$ (which is DC) $Z(0)$ = 2 kΩ.

Design this circuit with real components that can be readily obtained (i.e., with standard 5% resistors and available inductors and capacitors).

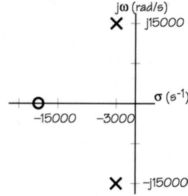

FIGURE 6.30: Desired $Z(s)$.

6.5.1   Try No. 1

Hmmm, we need a DC path through the circuit because we want $Z(0)$ = 2 kΩ. This says there must be a resistor across the terminals. We will also want both an inductor and a capacitor to be able to have complex poles. At the same time, we don't want the inductor across the terminals because it acts like a short circuit at DC. So let's consider the circuit of Fig. 6.31.

Compute $Z(s)$ for this:

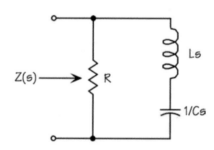

FIGURE 6.31: Possible circuit.

$$Z(s) = \frac{R\left(Ls + \dfrac{1}{Cs}\right)}{R + Ls + \dfrac{1}{Cs}} = \frac{R\left(s^2 + \dfrac{1}{LC}\right)}{s^2 + \dfrac{R}{L}s + \dfrac{1}{LC}}$$

Can we put numbers into this and fit the specified s-plane? Well, the denominator looks like it will be OK, because we should be able to find values that will give us complex roots. But the numerator won't do—it will produce a pair of complex zeros (at $\pm j\sqrt{1/LC}$)

Nice try, but….

6.5.2   Try No. 2

Let's make a DC path through the inductor. We still need both an inductor and a capacitor, but I'll move the inductor so that it's in series with the resistor. At DC the inductor is a short circuit, so the resistor is still across the terminals at DC. Fig. 6.32 shows this circuit.

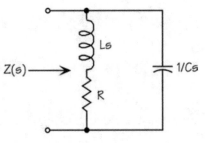

FIGURE 6.32: Another possible circuit.

Write $Z(s)$ for this circuit:

$$Z(s) = \frac{(Ls+R)\dfrac{1}{Cs}}{Ls+R+\dfrac{1}{Cs}} = \frac{\dfrac{1}{C}\left(s+\dfrac{R}{L}\right)}{s^2+\dfrac{R}{L}s+\dfrac{1}{LC}}$$

That looks pretty good. The denominator is the same as in the first try, and the numerator now has just a single zero.

To find numbers, I need to write $Z(s)$ for the function shown on the s-plane in Fig 6.30:

$$Z(s) = \frac{K(s+\alpha)}{(s+3000+j1\ 000)(s+3000-j1\ 000)}$$

$$= \frac{K(s+\alpha)}{s^2+6000s+234\times10^6}$$

(Note that $-\alpha$ is the position of the unspecified zero.)

Now I'll match the coefficients of the circuit's $Z(s)$ with the s-plane's $Z(s)$.

$$\frac{R}{L}=6000, \frac{1}{LC}=234\times10^6,$$

$$\alpha = \frac{R}{L}=6000$$

Oh, I haven't used the specification that $Z(s)$ at $s=0$ is to be 2 kΩ. From the circuit,

$$Z(0) = \frac{\frac{1}{C}\left(\frac{R}{L}\right)}{\frac{1}{LC}} = R$$

So R must be 2000. Solving these equations gives

$$R = 2000 = 2 \ \Omega,$$

$$\frac{R}{L} = 6000, L = \frac{R}{6000} = 0\ 333 \quad ,$$

$$\frac{1}{LC} = 234 \times 10^6, C = \frac{1}{234 \times 10^6 L} = 0\ 012 \ \mu$$

6.5.3 Commercial Values

Now I must reduce these values to values of devices I can purchase:

- The resistor R = 2 kΩ is a standard 5% value.
- An inductor of 330 mH is available commercially; JW Miller 70F331AF is one example.
- A capacitor of 0.012 μF is available; Panasonic ECQ-B1123JF, which has a working voltage of 100 V, is one.

Fig. 6.33 shows this circuit with the commercial values of the components.

But does this have the correct s-plane as specified at the outset? We need to calculate the positions of the singularities.

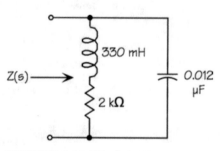

FIGURE 6.33: Final circuit.

$$Z(s) = \frac{\frac{1}{0\ 012 \times 10^{-6}}\left(s + \frac{2000}{0\ 33}\right)}{s^2 + \frac{2000}{0\ 33}s + \frac{1}{0\ 33 \times 0\ 012 \times 10^{-6}}}$$

$$= \frac{3\ 33 \times 10^6 (s + 6061)}{(s + 3030 + j1\ 600)(s + 3030 - j1\ 600)} \ \Omega$$

Fig. 6.34 shows the final s-plane.

Finally, check $Z(0)$ by substituting $s = 0$ into the final equation for $Z(s)$:

$$Z(0) = \frac{3\ 33 \times 10^6 (6061)}{1\big/\left(0\ 33 \times 0\ 012 \times 10^{-6}\right)} = 2000\ \Omega$$

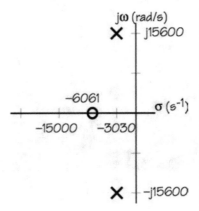

FIGURE 6.34: s-plane for final circuit.

Well! That all came out fine. The poles are off by 4% in the $j\omega$ direction and only 1% in the σ direction. The magnitude at $s = 0$ is correct. Finally, there is a single zero at $s = -6061$, not at the origin.

The circuit meets specification.

6.6   SUMMARY

What have we done in this chapter? Worked with impedances and transfer functions that involve the complex frequency s. Learned a few things about characteristics of functions of s. Drawn the s-plane and interpreted it to describe that circuit performance.

Hmmm, not too much, is there? And yet something fundamental is hiding in all this. Functions of s can be represented on the s-plane. The s-plane gives us a feeling—insight, no less—for how a circuit is going to perform over a range of values of ω. We can know how a certain circuit can be expected to perform in the frequency domain.

What does this mean for us? If we need to design a filter to handle some aspect of separating parts of a signal in the frequency domain, we need to be able to work in that domain. An understanding of possible configurations of the s-plane can lead us to circuits that could do the job.

We have one more step to go. In the next chapter we will restrict our study of the response of systems in the frequency domain by allowing s to take on only imaginary values ($j\omega$). This brings us to the frequency-domain equivalent of the sinusoidal steady state. And that's where a lot of action is in electrical systems.

CHAPTER 7

Frequency Response: ω is King

Remember the *s*-plane? Where *s* is king? Where the domain of functions such as $Z(s)$ and $H(s)$ is the complex frequency $s = \sigma + j\omega$? Where we drew this plane to show where the functions did crazy things? And where we excited the systems with inputs of the form $A\,e^{-\sigma t}\cos\omega t$?

I wonder how many sources can produce inputs like that? Certainly not in the power system, where the input is typically of the form $A\cos 377t$ (at least in North America). Probably not in communications systems, where we break signals down into sinusoidal components.

Consider communications for a moment. As engineers we are usually not concerned with the message content of the signal we are working with. Instead, we are concerned with the signal itself, and in particular, its frequency content. For example, the telephone system generally passes signals whose frequencies are between about 300 and 3400 Hz. A stereophile will want an amplifier/speaker combination to pass signals from below 20 Hz to at least 20 kHz, all at the correct amplitude and phase.

Note that everything in the preceding paragraph is stated in terms of "frequency." In this chapter we are going to look in general at the "frequency response" of systems. In other words, we'll look more at $H(\omega)$ than at $H(s)$. And since we are generally used to talking about "cycle frequency" in hertz rather than radian frequency in radians per second, we'll often shift from ω (rad/s) to f (Hz) ($\omega = 2\pi f$).

Yet this all still relates to the *s*-plane, as we will see shortly, but with a restricted view. In most cases, we don't want the transient behavior of the system when we are looking at frequency response. Instead, we want the results in the *sinusoidal steady state*, so we will learn to find the sinusoidal steady-state time-domain response directly without using inverse transforms.

One common way of presenting the sinusoidal steady-state response in the frequency domain is through *Bode diagrams*. These sound hairy, but you've probably already seen them if you read anything in, say, audiophile magazines or ham radio literature. We'll look at modern (computer age) uses of these graphs.

Finally, we'll take a look at one topic that will serve to bring together a number of things we've been learning—resonance. It's a useful topic in its own right, and it's important in systems, too.

7.1 H(jω)

A simple way to start all this is to take a look at an example that will show where we are going. Consider the transfer function $H(s)$:

$$H(s) = 1000\frac{s}{s^2 + 100s + 1002500}$$

$$= \frac{1000s}{(s+50+j1000)(s+50-j1000)}$$

Let's plot in Fig. 7.1 the three-dimensional surface that $H(s)$ represents.

Now let's look at the s-plane for this, shown in Fig. 7.2:

So far, so good. We can clearly see that there are two poles, one at $-50 + j1000$ and the other at $-50 - j1000$ rad/s. These are pretty obvious in the 3D plot, too. And there's a zero at the origin, which is a little trickier to spot in the 3D plot.

OK, now let's compute the magnitude of $H(s)$ at a point *on* the *jω* axis. I'll do this on the *s*-plane as shown in Fig. 7.3, knowing that the scale factor for this *s*-plane is 10^3.

Using just the magnitudes of the vectors shown on the diagram, we'll get

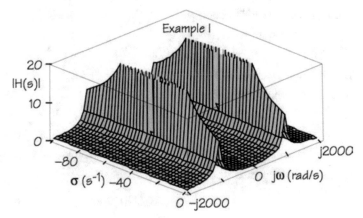

FIGURE 7.1: 3D version of $H(s)$.

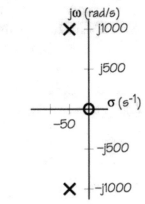

FIGURE 7.2: s-plane version of $H(s)$.

$$|H(j\omega_1)| = \frac{1000\omega_1}{\sqrt{(\omega_1 + 1000)^2 + 50^2}\sqrt{(1000-\omega_1)^2 + 50^2}}$$

Messy? Sure, but we have computers to help us. Let's see what happens if we plot this function for ω_1 running from −2000 to +2000 rad/s. Figure 7.4 shows this.

Do you see anything special? Sure, a couple of peaks, very close to where the poles are on the s-plane. And a zero at the origin. But note something else by comparing this plot with the front right face of the 3D plot of Fig. 7.1. If we had restricted the 3D plot of Fig. 7.1 to just the front right face, that plot would look just like the 2D plot of Fig. 7.4. To say it another way, the ω plot, which includes just the $j\omega$ axis, is a slice out of the 3D diagram of $H(s)$.

There's an important point here. The "frequency response" of a system described by $H(s)$ is the vertical slice of the 3D plot that includes the $j\omega$ axis. In other words, we are restricting our attention to $s = j\omega$, an s with no real part.

We will do a lot with frequency response before this chapter ends!

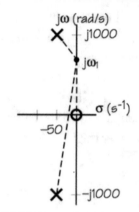

FIGURE 7.3: Calculation of $|H(j\omega_1)|$.

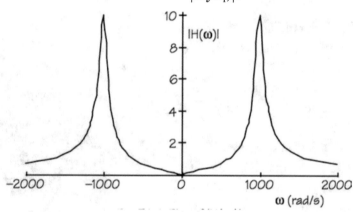

FIGURE 7.4: Plot of $|H(j\omega)|$.

7.2   SINUSOIDAL STEADY STATE

As I said at the beginning, we can get the sinusoidal steady-state time-domain response for a system directly from $H(s)$ without going through the inverse transform. But to develop this, I need to take us in an arm-waving manner through the inverse transform once.

Let's consider a system whose $H(s)$ is known and whose input is a sinusoid $v_{in}(t) = V_m \cos \omega_1 t\, u(t)$. Then we can write in the frequency domain

$$V_{in}(s) = V_m \frac{s}{s^2 + \omega_1^2},$$

$$V_{out}(s) = V_m \frac{s}{s^2 + \omega_1^2} H(s).$$

Suppose we want to get $v_{out}(t)$. We would do a partial-fraction expansion of $V_{out}(s)$ and then transform the terms back into the time domain. Avoiding details, I'll do this part way:

$$V_{out}(s) = \frac{As + B}{s^2 + \omega_1^2} + H_t(s).$$

Now comes the arm-waving argument. The first term of V_{out} is from the input to our system, and this term appears as one of the terms in the partial-fraction decomposition.

How about the second term, $H_t(s)$? This must be the rest of the partial-fraction decomposition. Since the first term comes from the input, the remaining term must come entirely from the original $H(s)$. Hence it must contain *all* of the poles of $H(s)$, and *only* those poles. In other words, the denominators of the rest of the partial-fraction result must be entirely from the denominator of $H(s)$.

If $H(s)$ is to be a real circuit, one that can be realized with resistors and capacitors and inductors, one that will be stable, the poles of $H(s)$ must be in the left half of the s-plane. In other words, they cannot have positive real parts. If they did, the system would be unstable and a small input would cause the output to head for infinity. (If the system's poles had zero real parts, the system would oscillate.)

Conclusion? All of the terms of the second part of our partial-fraction decomposition must be terms that will die out in the time domain. All of the parts of $H_t(s)$ must have poles only in the left-half plane, leading to time-domain exponentials with negative exponents.

Aha! If $H_t(s)$ contains only terms that die out, then the first term of V_{out} must represent the sinusoidal steady state. But we aren't done yet. Let's carry this decomposition a little farther by reducing the first term to two of lesser degree:

$$V_{out}(s) = \frac{C}{s - j\omega_1} + \frac{D}{s + j\omega_1} + H_t(s)$$
where $D = C^*$ (conjugate).

To find C, we multiply through by its denominator $(s - j\omega_1)$ and then let $s = j\omega_1$:

$$(s\text{-}j\omega_1)V_{out}(s) = \frac{C(s - j\omega_1)}{s - j\omega_1} + \frac{D(s - j\omega_1)}{s + j\omega_1} + (s - j\omega_1)H_t(s).$$

Before letting $s = j\omega_1$, I must make sure that there won't be a pole $s - j\omega_1$ in $Ht(s)$. If there were, it would cancel the $s - j\omega_1$ term in front of it. But there can't be such a pole in $Ht(s)$. If there were, $H(s)$ would have a pole *on* the $j\omega$ axis. Such a pole violates our restriction of having poles only *in* the left half-plane. (A pole *on* the $j\omega$ axis would produce a continuous output for a bounded input.)

With that argument, we can take the limit:

$$
\begin{aligned}
C &= \lim_{s \to j\omega_1} \left[(s - j\omega_1)V_{out}(s) - \frac{D(s - j\omega_1)}{s + j\omega_1} - (s - j\omega_1)H_t(s) \right] \\
&= \lim_{s \to j\omega_1} \left[(s - j\omega_1)V_m \frac{s}{s^2 + \omega_1^2} H(s) \right] \\
&= \lim_{s \to j\omega_1} \left[V_m \frac{s}{s + j\omega_1} H(s) \right] \\
&= V_m \frac{j\omega_1 H(j\omega_1)}{j\omega_1 + j\omega_1} = V_m \frac{H(j\omega_1)}{2}.
\end{aligned}
$$

To simplify things, let's note that $H(j\omega_1)$ is complex and hence has both magnitude and phase. I'll write the magnitude as simply $|H_1|$ and the phase as $\angle\theta_1$. Therefore

$$
C = V_m \frac{|H_1|}{2} \angle\theta_1.
$$

D is just the complex conjugate of C, so

$$
D = C^* = V_m \frac{|H_1|}{2} \angle -\theta_1.
$$

This gives us V_{out} at $s = j\omega_1$:

$$
V_{out}(j\omega_1) = \frac{V_m |H_1| \angle\theta_1}{2(s - j\omega_1)} + \frac{V_m |H_1| \angle -\theta_1}{2(s + j\omega_1)} + \text{dying terms}.
$$

Now ignore the dying terms and return to the time domain through the inverse transform, not forgetting to take the real part of the result. (I'm skipping a few of the algebraic steps that include the use of Euler's formula.)

$$
\begin{aligned}
v_{out\ ss}(t) &= \text{Re}\left[\left((V_m/2)|H_1| \angle\theta_1 \right) e^{j\omega_1 t} + \left((V_m/2)|H_1| \angle -\theta_1 \right) e^{-j\omega_1 t} \right] \\
&= |H_1| V_m \cos(\omega_1 t + \angle\theta_1).
\end{aligned}
$$

The result of all this effort is hidden away in the last equation. This says, if we have a system described by $H(s)$ and we want the sinusoidal steady-state output for a particular input frequency ω_1, we should

- evaluate $H(s)$ for $s = j\omega_1$, where ω_1 is the applied frequency,
- multiply the amplitude of the input by $|H(j\omega_1)|$ to get the amplitude of the output, and
- add the phase angle of $H(j\omega_1)$ to the phase angle of the input to get the phase angle of the output.

Uhuh, sure…show me an example. OK! Here's an example using the $H(s)$ from the beginning of this chapter and applying a cosine input. We want $V_{out}(t)$ in the steady state.

$$H(s) = 1000 \frac{s}{s^2 + 100s + 1002500},$$
$$v_{in}(t) = 100 \cos(500t + 45°) \text{ V}.$$

First, evaluate $H(s)$ at the input frequency $\omega = 500$ rad/s:

$$H(j500) = 1000 \frac{j500}{(j500)^2 + 100(j500) + 1002500}$$
$$= \frac{500000\angle 90°}{754159\angle 3.8°} = 0.663\angle 86.2°.$$

Now write the time domain result:

$$v_{out\,ss}(t) = 0.663 \times 100 \cos(500t + 45° + 86.2°)$$
$$= 66.3 \cos(500t + 131.2°) \text{ V}.$$

Just so that we don't forget where we came from, let's look back at the s-plane for this transfer function and note that we can "read" a result from the s-plane directly. (We know that the scale factor for this s-plane is 1000.)

Let's use the s-plane to find the output for the same input of $10 \cos(500t + 45°)$ V. Fig. 7.5 shows the s-plane for $H(s)$ with vectors drawn on it for $s = j500$ rad/s.

Since the vectors are either vertical, horizontal, or very close to vertical, we can approximate their angles and lengths:

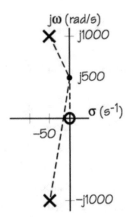

FIGURE 7.5: s-plane for $H(j500)$.

- From the zero, the length is 500 and the angle is 90°.
- From the upper pole, the length is close to 500 and the angle is almost −90°. (The triangle formed by the vector and the axis has sides of 50 and 500, so exact numbers are 502.5 and −84.3°.)
- From the lower pole, the length is close to 1500 and the angle is almost 90°. (The triangle has sides of 1500 and 50.)

Now, knowing the scale factor, we can write

$$H(j1000) \approx 1000 \frac{500 \angle 90°}{500 \angle -90° \times 1500 \angle 90°}$$
$$\approx 0.667 \angle 90°.$$

Combining this with the time-domain input gives us an approximate result that's very close to the formal result:

$$v_{out\,ss}(t) = 0.667 \times 100 \cos(500t + 45° + 90°)$$
$$= 66.7 \cos(1000t + 135°)\,\text{V}.$$

Two things are worth noting:

1. The s-plane carries all the information we need about a function of s except the scale factor. Hence we can get a feeling for what is happening in a system by looking at the s-plane and imagining the vectors to a point ω_1 on the $j\omega$ axis. Then we can visualize what happens as ω_1 varies along that axis.
2. This circuit exhibits a phenomenon called resonance. The output peaks at a frequency of about 1000 rad/s (i.e., at a frequency ω on the $j\omega$ axis in the vicinity of the pole). That peak is ten times as large as the input! We'll take up resonance in Section 7.4.

7.3   BODE DIAGRAMS

An American engineer, Hendrick Bode (*Boh' duh* with a long o although often mispronounced *Boh' dee*) developed a method for plotting the magnitude and the phase of $H(s)$ as a function of frequency (generally $f = \omega/2\pi$). He did some very clever things to make the job quite easy, at least for his time, because he found ways to get a decent plot without using a computer. (Which was smart on his part because there weren't any computers in the 1930s.)

The general idea is that the frequency response (i.e., the magnitude and the phase of a function of ω) can be determined by examining the poles and zeros of the function. Today, with computers all over the place, the details aren't too valuable. But Bode's name is still attached to the manner of presentation.

A *Bode diagram* is a graph of the base-10 log of the magnitude response of a system versus the base-10 log of the applied frequency, or a graph of the phase angle of the system's response versus the log of the applied frequency.[1]

Bode diagrams are common in some things that you may have seen. For example, the frequency response of a loudspeaker is generally shown as a Bode diagram.

7.3.1   Decibels

Logarithmic ratios are used in lots of places. Strengths of earthquakes is one example—the Richter Scale is logarithmic. In electrical systems we often use logarithmic ratios because we want to cover a wide range of a variable such as frequency.

The *decibel* (dB) is such a logarithmic ratio. The *bel* part of the word honors Alexander Graham Bell (or perhaps on 75% of him). The *deci* part is one-tenth. So what is this decibel? It was first a measure of sound levels.

The sound scale set 0 dB to be the sound level (defined in terms of sound pressure) that is supposed to be the tiniest sound detectable by the human ear (and then only if you are young and haven't listened to loud music). 1 dB is about the smallest *change* in sound intensity that our ears can detect. By using tenths of bels, a sound intensity of 100 dB is at about the painful level for us.

Note that a logarithmic scale requires a reference other than zero (the logarithm of zero is a bit of a disaster on a numerical scale). So whenever we talk about decibels, we need a starting point. For sound intensity, for example, this reference level is 10^{-16} W/cm^2.

Decibels first made their appearance in electrical systems as power ratios:

$$dB = 10 \log_{10} \frac{P_2}{P_1},$$

where the abbreviation dB represents decibels (the B is always capitalized). In this expression, P_1 could be the power into a certain circuit, while P_2 is the power out.

Note that this is a ratio, so the reference is the power input and the result is the gain of the circuit in dB.

Sometimes we want to talk about an absolute level in dB rather than being restricted to a ratio. In the communications industry, dB is often referenced to 1 mW. To distinguish this from ordinary dB, the quantity is called dBm, meaning "dB referenced to 1 mW."

---

[1]Bode actually developed his technique to plot the asymptotes of the frequency response and then fill in the actual curve by following simple algorithms.

We often talk about voltages and their ratios, and dB would be convenient for this. But to use dB properly for voltage ratios, we need a little different approach.

dB is primarily a power ratio. Remember, though, that power is voltage squared divided by resistance. So if we establish a certain value of resistance as our standard, we can relate power and voltage. Let's call our reference resistance R_{REF} and see what happens:

$$dB = 10 \log_{10} \frac{P_2}{P_1},$$

$$P_1 = V_1^2 / R_{REF}, P_2 = V_2^2 / R_{REF},$$

$$dB = 10 \log_{10} \frac{V_2^2 / R_{REF}}{V_1^2 / R_{REF}} = 10 \log_{10} \frac{V_2^2}{V_1^2}$$

$$= 10 \log_{10} \left(\frac{V_2}{V_1} \right)^2 = 20 \log_{10} \frac{V_2}{V_1}.$$

Now we can use dB for voltages, but the factor in front is now 20, not 10.

We still need a voltage reference (V_1). If V_1 is the input voltage to a circuit and V_2 is the output, then the voltage gain of the circuit is 20 times the base-10 log of the voltage ratio.

When it comes to the sinusoidal steady-state, we need to keep in mind that we are dealing with voltage *magnitudes* when we talk about dB. This means, then, that we can relate dB and the transfer function of a system:

$$dB = 20 \log_{10} |H(j\omega)|.$$

For communications applications, the dBm can also be used for voltages. The power reference, as I said before, is 1 mW. The reference resistance commonly used is 600 Ω. So you'll sometimes see a voltmeter with a dB scale and a notation, "1 mW into 600 Ω."

(There are other specialized forms of dB:

dBi   For comparing the power radiated by an antenna to that which would be radiated by an *isotropic* radiator connected to the same source.

dBd   Similar to dBi, but compared to a dipole.

dBW Power referenced to 1 W.

dBV   Voltage referenced to 1 V.

dBμV Voltage referenced to 1 μV.)

Because decibels are logarithmic, they add in cases where gains would multiply. For example, if we have an amplifier with a gain of 20 dB feeding a 10-dB attenuator, which is coupled by a patch panel with 0.1-dB loss to an amplifier with a gain of 26 dB, the overall gain of the system is $20 - 10 - 0.1 + 26 = 35.9$ dB.

We will use voltage dB heavily in Bode diagrams.

7.3.2   Example

Let's do an example. Here's that same $H(s)$ I've used throughout the chapter, with the s-plane for it shown in Fig. 7.6:

$$H(s) = 1000 \frac{s}{s^2 + 100s + 1002500}.$$

I used the computer to find the magnitude of $H(j\omega)$, convert this to dB, and plot dB versus radian frequency. The result is the Bode diagram in Fig. 7.7.

There's nothing terribly magical here (except perhaps figuring out how to make Mathematica or Maple give you axes with the correct labels).

Then I computed the phase angle for $H(j\omega)$ and plotted that versus ω as shown in Fig. 7.8.

Well, what do we see? Look at the peak on the magnitude plot; it appears to reach 20 dB at 1000 rad/s. 20 dB is a factor of 10, because 20 log 10 = 20 dB. Also note that the phase angle appears to pass through 0° at 1000 rad/s.

Do our results check at 500 rad/s? I used a square piece of paper and a pencil and read from the magnitude plot −4 dB and from the angle plot +86°. Checking −4 dB gives

$$-4\,dB = 20 \log_{10} x,$$

$$x = 10^{-4/20} = 0.631.$$

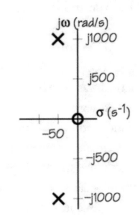

FIGURE 7.6: s-plane for Example I.

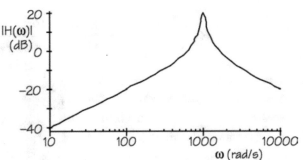

FIGURE 7.7: Bode diagram for $|H(j\omega)|$.

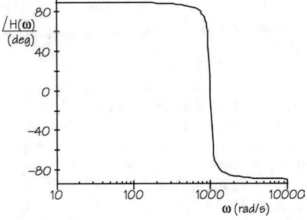

FIGURE 7.8: Bode diagram for $/H(j\omega)$.

The original calculations at the end of Section 7.2 gave 0.663 /86.2°. Not too bad.

More commonly, though, we use the log of "cycle frequency" in hertz as the horizontal axis. Let's do another example.

7.3.3   An Example in Hertz

Recall in Section 6.4 that we did a design of a circuit that was to have poles at −3000 ± j15,000 rad/s, a zero somewhere on the real axis in the left-half plane, and a DC impedance of 2 kΩ. The resultant circuit is shown in Fig. 7.9.

Let's use Bode diagrams to see how our circuit's impedance changes with frequency. This time, I'll replace jω with j2πf so that the results will be in hertz rather than in radians per second.

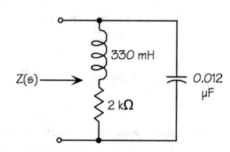

FIGURE 7.9: Final circuit of Section 6.4.

$$Z(f) = \frac{833.33 \times 10^6 (j2\pi f + 6061)}{(j2\pi f + 3030 + j15600)(j2\pi f + 3030 - j15600)} \; \Omega.$$

My plot of magnitude will use the base-10 log of that magnitude. Fig. 7.10 shows the magnitude plot, and Fig. 7.11 shows the phase plot. Both use the "hertz" calibration of the log-frequency axis.

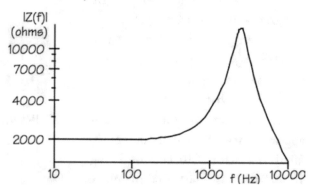

FIGURE 7.10: $|Z(f)|$ for Fig. 7.9.

Any news here?   Well, the impedance at DC (the diagram goes down only to 10 Hz) appears to be heading for 2000 Ω with a phase angle of 0°. There's a peak in the impedance of about 14 kΩ at about 2500 Hz, and the phase angle makes a large change there, too. This squares with the fact that there is a pole at −3030 + j15,600 rad/s = −482 + j2483 Hz. Therefore there should be a big change around 2500 Hz.

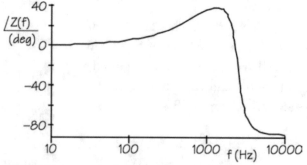

FIGURE 7.11: $/Z(f)|$ for Fig. 7.9.

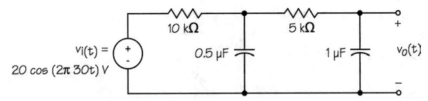

FIGURE 7.12: Circuit for Example III.

7.3.4 Another Example

Here's another example. Consider the circuit shown in Fig. 7.12. Find the sinusoidal steady-state response for the input shown, and plot the Bode diagrams (magnitude and phase) for frequencies from 1 to 1000 Hz.

I used proportionality and Maple to find the transfer function:

$$H(s) = 40000 \frac{1}{s^2 + 800s + 40000}.$$

Evaluating this at $f = 30$ Hz ($\omega = 2\pi 30$ rad/s) gives

$$|H(2\pi 30)| = 0.265,$$
$$\angle H(2\pi 30) = -88.3°.$$

Writing the steady-state time-domain expression by inspection:

$$v_o(t) = 0.265 \times 20\cos(2\pi 30t - 88.3°)$$
$$= 5.30\cos(2\pi 30t - 88.3°) \text{ V}.$$

The Bode diagrams for the transfer function are shown in Figs. 7.13 and 7.14. They aren't too exciting!

Note that the output decreases with frequency. That's because the capacitors become more and more like short circuits as the frequency increases.

It's interesting to note that the ultimate downward slope of the magnitude curve is −40 dB per decade, where a

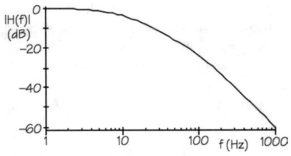

FIGURE 7.13: Magnitude diagram for Example III.

decade is a factor of 10 along the horizontal axis. Circuits made with real components all exhibit slopes that are integer multiples of 20 dB per decade.

In the phase plot, the starting angle is about 0° (and would be if we had extended the graph to the left). The ending angle is −180° (if we had gone far enough).

Mr. Bode (who died in 1982) would be delighted with the computer capability we have today! It would have simplified very greatly his book that he published in 1945 on systems and stability.

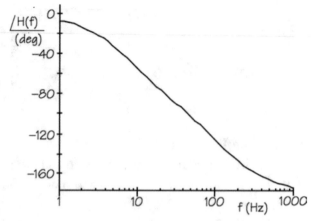

FIGURE 7.14: Phase diagram for Example III.

7.4 RESONANCE

We've already noted that we get a peaked response with some of our examples. This peaking is generally resonance, a phenomenon that is important in its own right.

Many systems exhibit a vigorous response at certain frequencies. You may have noticed this while bouncing on a bed or sloshing water in a pool: if you did things at just the right repetition rate you got a big response. You've also heard it in an auditorium when just the right energy was fed back from the hall to the microphone: the whole system resonated with a loud howl.

7.4.1 Series RLC Circuit

There are two important circuit configurations that exhibit resonance, both involving one inductor and one capacitor. (A system must be at least second order to exhibit resonance.) Let's consider first the series RLC circuit shown in Fig. 7.15.

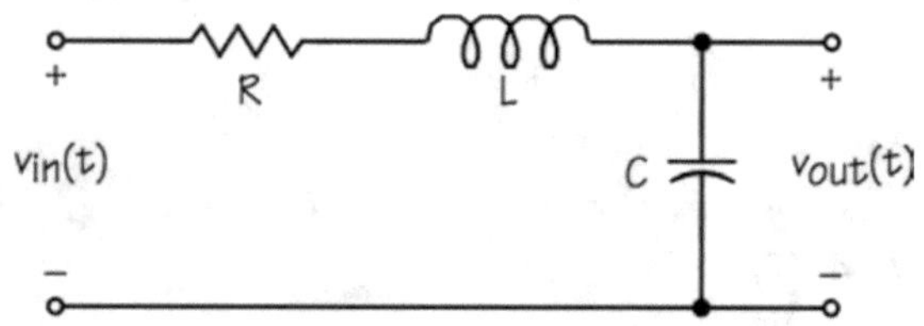

FIGURE 7.15: Series RLC circuit..

Let's work in the s-domain and calculate $H(s)$:

$$H(s) = \frac{\dfrac{1}{LC}}{s^2 + \dfrac{R}{L}s + \dfrac{1}{LC}}.$$

We are interested in the sinusoidal steady-state response, so I'll substitute $s = j\omega$.

$$H(j\omega) = \frac{\dfrac{1}{LC}}{(j\omega)^2 + \dfrac{R}{L} j\omega + \dfrac{1}{LC}}$$

$$= \frac{\dfrac{1}{LC}}{\left(\dfrac{1}{LC} - \omega^2\right) + j\dfrac{R}{L}\omega}.$$

What value of ω (call it ω_o) will make $|H(j\omega)|$ a maximum? It sure looks like $\omega_o^2 = 1/LC$, so

$$H(j\omega_o) = \frac{\dfrac{1}{LC}}{\left(\dfrac{1}{LC} - \omega_o^2\right) + j\dfrac{R}{L}\omega_o}$$

$$= \frac{\dfrac{1}{LC}}{(0) + j\dfrac{R}{L}\omega_o} = -j\frac{L}{R\sqrt{LC}}$$

$$= -j\frac{1}{R}\sqrt{\frac{L}{C}}.$$

Hence $|H(j\omega_o)| = \dfrac{1}{R}\sqrt{\dfrac{L}{C}}$ and $\angle H(j\omega_o) = -90°$.

Hmmm, the frequency (ω_o) for this maximum is determined by the inductor and the capacitor.

7.4.2 Parallel RLC Circuit

Let's try this again for the parallel RLC circuit. This is shown in Fig. 7.16. ("Parallel" and "series" refer to the arrangement of the inductor and the capacitor only.)

Making the same calculations,

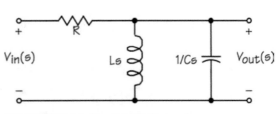

FIGURE 7.16: Parallel RLC circuit.

$$H(s) = \frac{V_{out}(s)}{V_{in}(s)} = \frac{\dfrac{L}{C}}{\dfrac{L}{C} + R\left(Ls + \dfrac{1}{Cs}\right)},$$

$$H(j\omega) = \frac{\dfrac{L}{C}}{\dfrac{L}{C} + R\left(j\omega L - j\dfrac{1}{\omega C}\right)}.$$

Again it appears that the magnitude of $H(j\omega)$ will be a maximum when

$$\omega_o L = \frac{1}{\omega_o C}, \omega_o = \frac{1}{\sqrt{LC}}.$$

7.4.3   Resonance and Bandwidth

But we need a more general approach to resonance. Both of the functions we've just dealt with have a quadratic denominator:

$$s^2 + As + B.$$

The numerator may be Ks or just K, but this isn't important here. Let's put all this together:

$$H(s) = \frac{numerator}{s^2 + As + B},$$

$$H(j\omega) = \frac{numerator}{(j\omega)^2 + A(j\omega) + B}$$

$$|H(j\omega)| = \frac{|numerator|}{\sqrt{(B - \omega^2)^2 + A^2\omega^2}}.$$

At resonance, the denominator term in parentheses becomes zero, so

$$B - \omega_o^2 = 0, \omega_o = \sqrt{B}, \text{ so}$$

$$|H(j\omega)| = \frac{|numerator|}{\sqrt{(0)^2 + A^2\omega_o^2}} = \frac{|numerator|}{A\omega_o}.$$

That says that the resonant frequency is determined by the B of the denominator quadratic (provided the coefficient of s^2 is 1). This is the frequency at which the denominator is purely imaginary. It is also the frequency at which the magnitude of the response is a maximum.

We often talk about the "bandwidth" of a circuit (such as a filter). What we mean is the range of frequencies that will pass through the circuit. But we need something to standardize the way we establish bandwidth. We commonly use the "half-power points" on the magnitude version of the Bode diagram.

Why half power? Why not? It's as convenient as any other number. We could have used one-tenth, for example, but half is the usual one. The confusion comes when we have a voltage plot. We want half *power*. Power depends on voltage squared. So if *voltage* is reduced by the square-root of 2, that will be half *power*.

Here's a resonance plot (Fig. 7.17) with the half-power points marked. Since this is a voltage plot, I've used the $1/\sqrt{2}$ point (0.707).

Let's figure out what ω_1 in that plot must be for the general function we have been working with. I'll start with our original function and equate it to the peak value divided by the square-root of 2:

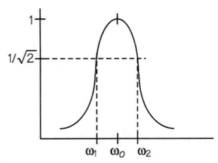

FIGURE 7.17: Half-power points.

$$|H(j\omega_1)| = \frac{|numerator|}{\sqrt{(B-\omega_1^2)^2 + A^2\omega_1^2}} = \frac{|numerator|}{\sqrt{2}A\omega_o}$$

For this to be true, the two terms inside the square root must be equal:

$$(B-\omega_1^2)^2 = A^2\omega_1^2,$$

$$\sqrt{(B-\omega_1^2)^2 + A^2\omega_1^2} = \sqrt{2A^2\omega_1^2} = \sqrt{2}A\omega_1,$$

$$B-\omega_1^2 = \pm A\omega_1,$$

$$\omega_1^2 \pm A\omega_1 - B = 0,$$

$$\omega_1 = \pm\frac{A}{2} \pm \sqrt{\left(\frac{A}{2}\right)^2 + B}.$$

All the ± signs come from remembering that a square root can be either + or −. Only one of the four possible arrangements of signs will produce a positive frequency.

If we do the same thing for ω_2, we'll get a similar result. Keeping only the positive-frequency terms gives

$$\omega_1 = -\frac{A}{2} + \sqrt{\left(\frac{A}{2}\right)^2 + B},$$

$$\omega_2 = +\frac{A}{2} + \sqrt{\left(\frac{A}{2}\right)^2 + B}.$$

Bandwidth is defined as the distance between the half-power points, so

$$BW = \omega_2 - \omega_1$$

$$= \left[+\frac{A}{2} + \sqrt{\left(\frac{A}{2}\right)^2 + B}\right] - \left[-\frac{A}{2} + \sqrt{\left(\frac{A}{2}\right)^2 + B}\right]$$

$$= A.$$

So the bandwidth is just A! You might consider trying the product of ω_1 and ω_2 and seeing what you get.

7.4.4   Q

The "quality factor" tells how sharp the peak is, or how narrow the bandwidth is. We define this quality factor Q as

$$Q = \frac{\omega_o}{BW} = \frac{\omega_o}{A}.$$

A Q of 5 is often considered the dividing line between low- and high-Q circuits (i.e., wide bandwidth or narrow bandwidth). A Q of more than 5 is considered a high-Q circuit with a narrow bandwidth.

We can also derive various forms of Q for parallel and series RLC circuits in terms of the circuit components, but we won't do that here.

Q is also defined for inductors alone:

$$Q = \frac{\omega L}{R},$$

where ω is the operating frequency, L is the inductance, and R is the equivalent series resistance of the coil.

7.4.5 Resonance Example

Let's finish this with a somewhat different example. While I've derived all of this stuff using the parallel or the series RLC circuit, we don't have to stick with just those combinations. As long as we have at least a second-order system (at least a quadratic in the denominator of $H(s)$), we can have resonance.

The circuit of Fig. 7.18 can be made to exhibit resonance. I have chosen circuit values so this is true. Find the resonant frequency, the bandwidth, and Q for this.

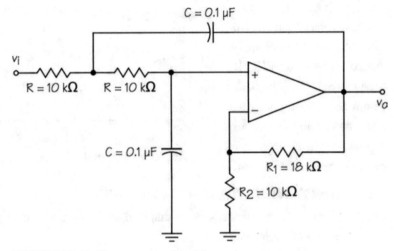

FIGURE 7.18: Resonant circuit with op-amp.

I used the computer to do the analysis in the frequency domain:

$$H(s) = \frac{2800000}{s^2 + 200s + 1000000},$$
$$A = 200, B = 1000000,$$
$$\omega_o = \sqrt{B} = 1000 \text{ rad/s},$$
$$\omega_1 = -\frac{A}{2} + \sqrt{\left(\frac{A}{2}\right)^2 + B} = 905 \text{ rad/s},$$
$$\omega_2 = +\frac{A}{2} + \sqrt{\left(\frac{A}{2}\right)^2 + B} = 1105 \text{ rad/s},$$
$$BW = \omega_2 - \omega_1 = 200 \text{ rad/s},$$
$$Q = \frac{\omega_o}{BW} = 5.$$

It's interesting to look at the Bode diagrams for this. These are shown in Figs. 7.19 and 7.20.

Do you see reso-
nance? Note the peak
and the sharp phase-angle
change. This circuit does
not have a very high Q. We
can show that the Q here is
controlled by R_1 and R_2. If
the Q were higher, the peak
would be sharper, giving us
a narrower bandwidth and
a phase-angle change that
would be even more abrupt.

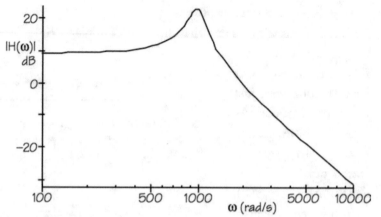

FIGURE 7.19: Magnitude diagram for Example IV.

7.5   ANOTHER EXAMPLE OR TWO

More s-plane stuff, resonance, a Bode diagram, and a design from a Bode diagram are covered
in the examples that follow.

7.5.1   Example I

The circuit of Fig. 7.21
will be used in this
example and the next
three as well. In this
example, I want the
impedance seen from
the input terminals.
Sketch the s-plane for
this impedance.

This is a straight-
forward problem—find
the impedance of each

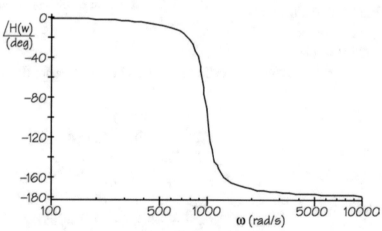

FIGURE 7.20: Phase diagram for Example IV.

of the elements, put 'em in series, and simplify the result so it's in standard form:

$$R = 250\,\Omega,$$
$$Z_L = 200 \times 10^{-3} s\,\Omega,$$
$$Z_C = 1/0.04 \times 10^{-6} s\,\Omega,$$
$$Z(s) = R + Z_L + Z_C = 0.2\frac{s^2 + 1250s + 125 \times 10^6}{s}\,\Omega.$$

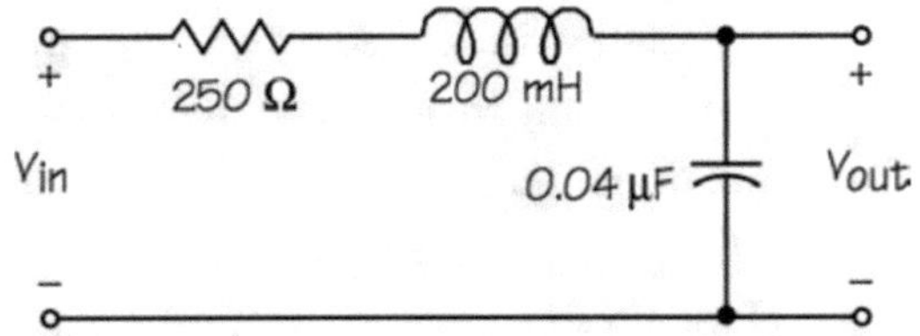

The standard form for presenting an impedance is as the ratio of two polynomials in s with 1 as the coefficient of the highest power of s in each. (There is another common form that has 1 as the coefficient of the lowest power of s in each—controls folks usually use this.)

For the s-plane I need the poles and zeros:

$$Z_{zeros} = -625 \pm j11163\,\text{s}^{-1},$$
$$Z_{pole} = 0.$$

Fig. 7.22 is a completely labeled sketch of the s-plane for this example.

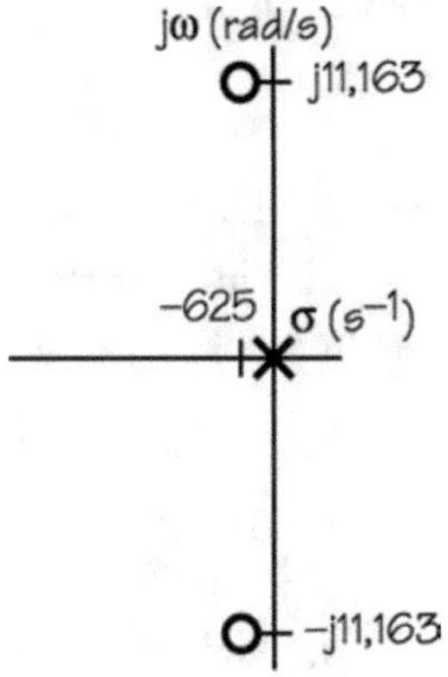

7.5.2 Example II

For the same circuit as in Example I, find the voltage transfer function $H(s) = V_{out}/V_{in}$. Then sketch the s-plane for $H(s)$.

This is a voltage divider, so $H(s)$ is

$$H(s) = \frac{Z_C}{Z_C + Z_L + R} = 125 \times 10^6 \frac{1}{s^2 + 1250s + 125 \times 10^6}.$$

$H(s)$ has no zeros; the poles are

$$H_{poles} = -625 \pm j11163\,\text{s}^{-1}.$$

The s-plane is shown in Fig. 7.23.

7.5.3 Example III

For the $H(s)$ found in the previous example, find the bandwidth, the resonant frequency, and the Q of the circuit.

If the denominator polynomial of $H(s)$ is in standard form, i.e., with 1 as the coefficient of the s^2 term, the bandwidth and the resonant frequency can be read from the remaining coefficients.

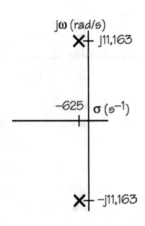

FIGURE 7.21: Circuit for Example I, II, III, and IV.

250 Ω 200 mH 0.04 μF V_{in} V_{out}

$j\omega$ (rad/s)

$j11,163$

-625 $\sigma\,(\text{s}^{-1})$

$-j11,163$

FIGURE 7.22: Example I: $Z(s)$.

$j\omega$ (rad/s)

$j11,163$

-625 $\sigma\,(\text{s}^{-1})$

$-j11,163$

FIGURE 7.23: Example II: $H(s)$.

The bandwidth is the coefficient of the s^1 term, so here the bandwidth is 1250 rad/s.

The resonant frequency is the square root of the coefficient of the s^0 term. Here, this frequency, ω_o, is the square root of 125×10^6, which is 11,180 rad/s. In hertz, f_o is 1780 Hz. (Divide the radian frequency by 2π.)

Q for a resonant circuit is the resonant frequency divided by the bandwidth (in consistent units). Here, $Q = 11{,}180/1250 = 8.94$. (That's a fairly high Q and, as we'll see in the next example, makes a neat peak on the Bode diagram.)

7.5.4   Example IV

Sketch the Bode diagram for the $H(s)$ of the previous examples.

The Bode diagram is, as understood by most engineers, a plot of the magnitude of the "gain" of the system versus the frequency in hertz. The magnitude of the gain is given in dB and the frequency axis is logarithmic, so the Bode plot is really a log-log plot.

I'll just list the steps to create the plot. If you do this in Maple or Matlab, the printed results are messy and don't convey much information.

1. Replace s in H(s) with $j2\pi f$.
2. Compute the magnitude of the resultant H(f).
3. Compute $20 \log_{10} |H(f)|$ to get the result in dB.
4. Make a semilog plot of dB versus $\log_{10} f$.

The Bode diagram that results in Fig. 7.24.

7.5.5   Example V

Design a circuit that has an $H(s)$ that will yield the Bode plot shown in Fig. 7.25. Keep the circuit as simple as possible. Use standard commercial components.

(The diagram is plotted using ω rather than f for convenience. If the plot had been with f, I'd have to be smart enough to remember to multiply all frequencies by 2π.)

What can I read from the diagram?

1. The plot starts with an upward slope of 20 dB/decade. Hence it begins "a long way down and a long way to the left." In other words, $H(s)$ is 0 at DC, which means there is no DC path through the circuit.

2. In the middle of the graph, the downward break of the slope from 20 dB/decade upward to 0 dB/decade indicates that there is one pole on the negative real axis to cause this break.

3. The high-frequency response of the circuit is flat, which means there are no more real-axis poles or zeros.

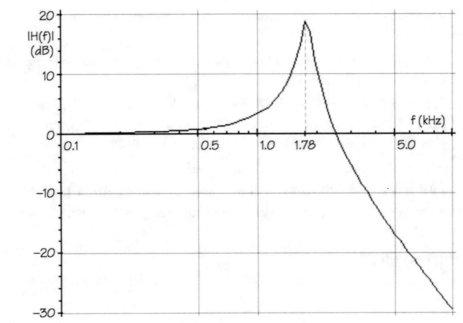

FIGURE 7.24: Example II: Bode diagram.

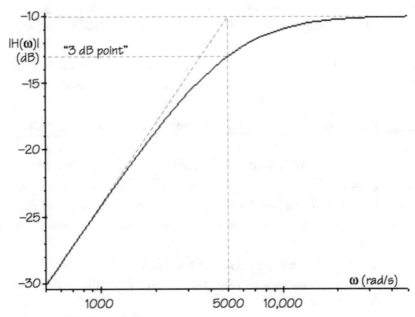

FIGURE 7.25: Example V: design goal.

4. The location of the pole can be found by either of two methods. Both are shown on the graph:

   a. Measure 3 dB downward from the flat part of the curve and note the frequency. On the graph this is 5 krad/s.

   b. Sketch in the asymptote to the 20 dB/decade slope. (Be sure to follow the "really straight" part of the slope.) Also sketch the asymptote of the flat part of the curve. Note where the asymptotes intersect. Again, it's 5 krad/s.

   This places the lone pole at −5000 s⁻¹ on the *real* axis.

5. The flat area is at −10 dB, so the high-frequency "gain" of the circuit $10^{(-10/20)} = 0.316$.

6. The circuit is a high-pass filter with a cutoff frequency of 5 krad/s, which is about 800 Hz.

From all this, I can write $H(s)$:

$$H_{goal}(s) = \frac{Ks}{s + 5000},$$
$$K = 10^{(-10/20)} = 0.316.$$

Now I need to think a bit. What are the low-frequency (DC) and high-frequency characteristics of our usual circuit elements? Well, resistors, ideal ones, at least, aren't bothered by frequency. Capacitors are open circuits at DC and short circuits at the other end. Inductors are just the opposite.

I could place a capacitor in series with the input, which will make the DC response zero. I could place an inductor across the output and achieve the same result. Is one choice better than the other?

Inductors are expensive, often large, and always have an associated resistance that can be modeled as being in series with the inductor. Hence it will be hard to get a solid short circuit across the terminals. While that "short" could be a small resistance when compared with other resistors in the circuit, still, the inductor doesn't sound like a good choice.

Capacitors tend to be small, at least for lower voltages, and they tend to be rather stable. Many exhibit very little leakage, megohms, perhaps. So a capacitor in series with the input will be my choice.

Finally, I need a high-frequency "gain" of less than 1. This sounds like a voltage divider. Now I can draw the circuit and see whether it will have the correct form of H(s). The proposed circuit is shown in Fig. 7.26.

For this circuit, $H(s)$ is

$$H(s) = \frac{R_2}{R_2 + R_1 + 1/Cs} = \frac{\dfrac{R_2 s}{R_2 + R_1}}{s + \dfrac{1}{R_2 C + R_1 C}}.$$

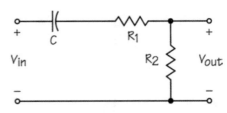

Now just match coefficients:

FIGURE 7.26: Example V: trial circuit.

$$\frac{R_2}{R_2 + R_1} = 0.316, \quad \frac{1}{R_2 C + R_1 C} = 5000.$$

This gives me two equations in three unknowns. Because commercial values of capacitors are more limited than those of resistors, I'll solve these for the resistors in terms of the capacitors:

$$R_1 = \frac{136.75}{C}, R_2 = \frac{63.246}{C} \text{(C in } \mu F).$$

Now it's a matter of choosing a value of C and checking the resistor values to see if they are practical. I will choose to keep the resistor values in the usual range for op-amp and signal circuits, namely, a "few k."

For capacitor values I tried first 1 μF, which made the resistors too small. I worked through 0.1 μF, 0.022 μF, and finally settled on 0.047 μF. This yielded the resistor values of R_1 = 2910 Ω, which becomes the 5% value of 3.0kΩ, and R_2 = 1346 Ω, which becomes 1.3 kΩ.

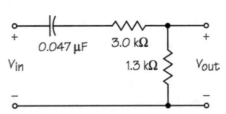

The final circuit is shown in Fig. 7.27. The Bode diagram of Fig. 7.28 plots the desired and actual curves together—the upper line is the desired response. The difference is less than 1 dB everywhere.

FIGURE 7.27: Example V: design result.

7.6 CIRCUIT DESIGN EXAMPLE

AM radio reception (the "common" band between 550 and 1600 kHz) is often bothered by a 10-kHz whistle. This is caused by the "beating" of the carrier of the desired station with a station on an adjacent frequency.

Many receivers have an "AM whistle filter" or a "beat cut" that reduces this tone to a level that isn't as bothersome. My last example is to design such a filter using only passive components.

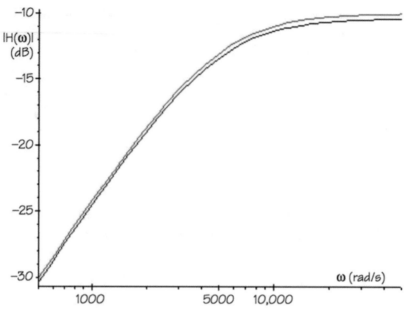

FIGURE 7.28: Example V: check of results.

7.6.1 Whistle Filter Specifications

The whistle filter is to be a "notch" at least 20 dB deep at 10 kHz. The bandwidth of the notch is to be no more than 1 kHz. Below the notch, the passband should be approximately flat from 20 Hz to 5 kHz. ("Flat" sometimes means "within 1 dB" and sometimes "within 3 dB." I'll use the poorer 3-dB definition because AM radio doesn't have too good a frequency response anyhow.)

This filter is to be part of a larger system (the receiver). The signal, having been received from the antenna, amplified, demodulated, and amplified some more, will come to our filter from circuitry that looks like a 600-Ω source.

Fig. 7.29 is a stylized Bode plot of the desired frequency response of this filter. Note that "notch" really means we want a notch-like shape. Note also that the plot is at 0 dB to the left of the notch. It's also at 0 dB to the right of the notch, but this isn't required in the specifications.

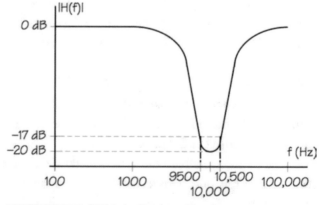

FIGURE 7.29: Whistle filter specifications.

What does "bandwidth…no more than 1 kHz" mean? That's the distance between the half-power points. But here, "half power" takes on a slightly different meaning. We are concerned with the width of the notch. So the half-power points are actually *above* the bottom of the notch. Since the bottom of the notch is at −20 dB, the half-power points are at −17 dB and their frequencies are 9.5 and 10.5 kHz.[2]

Oh oh, how did the half-power points fall at −17 dB? Recall that "half power" meant a voltage that is reduced by the square root of one half. Since this filter is upside down, the half-power points would be at double the power. That means the square root of double. In dB this becomes

$$20\log_{10}\left(\sqrt{2}/1\right) = 3 \text{ dB}.$$

So the half-power points that will define our maximum bandwidth of 1 kHz will be at −20 + 3 = −17 dB.

7.6.2   First Try

When we were looking at resonance, we noted that an inductor and a capacitor together can act to produce a peak in Bode diagrams of $Z(s)$ and $H(s)$. Recall the circuit shown in Fig. 7.30. Here the inductor and the capacitor are in parallel. (This is the same circuit we examined in Fig. 7.16.)

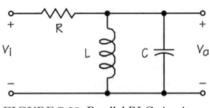

FIGURE 7.30: Parallel RLC circuit.

In this circuit, when the inductor and capacitor resonate, they act like an open circuit. This means that, at resonance, they appear not to be there, so the circuit is "straight through." The Bode diagram of the response of this circuit is shown in Fig. 7.31.

Of course, this is upside down from what we are trying to design. But it gives us a clue as to how to proceed. If the parallel LC combination at resonance is an open circuit, then it seems to make sense that the series combination at resonance would be a short

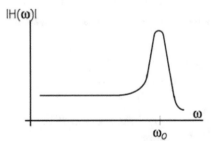

FIGURE 7.31: Bode diagram for parallel RLC.

[2]This isn't quite correct. The requirement that the half-power points be 1 kHz apart doesn't mean they should be at 9.5 and 10.5 kHz. They'll be close, however, almost centered on 10 kHz.

circuit. That would force the output to zero at the resonant frequency. The circuit is shown in Fig. 7.32.

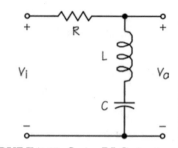

FIGURE 7.32: Series RLC circuit.

Now we need to see what the Bode diagram of this
circuit looks like. But that takes numbers. I'm going to
use a technique that is often used in circuit design, namely,
choosing all components to have values of 1. Once I know
how the circuit behaves, I can *scale* the circuit elements to
get the frequency response and the impedance level where
I want them. Fig. 7.33 is the Bode diagram for the "1"
circuit (R = 1 Ω, L = 1 H, C = 1 F).

Oh, wow! That is a steep, narrow notch! Too steep. It looks like it goes way down to
−50 dB. Actually, it goes down to −∞ dB because the series LC combination is a short circuit
at resonance. That's a little *too* good! And very unrealistic.

Note that the resonant frequency is

$$\omega_o = 1/\sqrt{LC} = 1\,\text{rad/s}.$$

7.6.3   Second Try

Too good, perhaps, and unrealistic.
It is impossible to make an inductor
that doesn't have resistance (except
with superconductors). So a real
inductor would have some resistance. Let's give the inductor some
resistance in our model and see
what happens. Fig. 7.34 is the new
circuit.

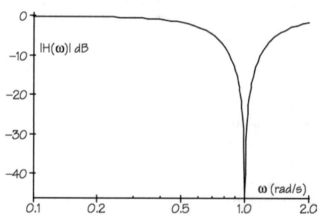

FIGURE 7.33: Frequency response of "1" circuit.

Now let's do the "1" thing
again. But I decided that making
R_L = 1 Ω would be too much resistance for an obvious reason: if the
L and C together at resonance are a
short circuit, the remaining circuit
is a voltage divider. This divider is R_L and R. So if I want $|H(\omega)|$ = −20 dB at resonance, then
R_L must be about one-tenth as large as R. (To be 100% correct, R_L should be one-ninth of R,
but I'll stick with simple numbers.)

Fig. 7.35 is the Bode diagram for my new circuit, using $L = 1$ H, $C = 1$ F, $R = 1\ \Omega$, and $R_L = 0.1\ \Omega$.

Gee, does that look good or what? Or what, actually, because I have really "lucked out." I need to analyze the circuit itself at four points:

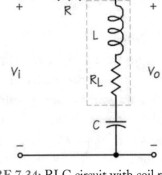

FIGURE 7.34: RLC circuit with coil resistance.

- The resonant frequency, ω_o (1 rad/s), which gives a result of −20.8 dB.
- The frequency $\omega_1 = 0.95$ rad/s, which is 95% of the resonant frequency. (I choose this because 9500 Hz is 95% of the desired 10,000 Hz.) At ω_1 the result is −17.7 dB.
- Similarly, at $\omega_2 = 1.05$ rad/s the result is −18.0 dB.
- An inspection of the graph shows the passband to be at 0 dB.
- At 0.5 rad/s (which corresponds to 5000 Hz in the final result), the graph shows the output to be about −2 dB.

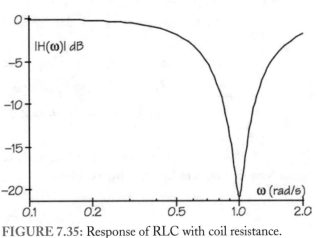

FIGURE 7.35: Response of RLC with coil resistance.

- The inductor is even reasonable. Its Q is important because it is expensive to make high-Q inductors. Recall that Q for an inductor is defined at the operating frequency:

$$Q = \frac{\omega L}{R_L} = \frac{1 \times 1}{0.1} = 10.$$

These all look fine! The notch is a little better than minus 20 dB. The half-power points aren't quite "half power" but they are close. At 0.95 we have 3.1 dB above −20.8, and at 1.05 we have 2.8 dB above −20.8. The passband looks good; it varies only about 2 dB from being flat where required. Finally, the inductor has a Q of 10, which we can probably build.

But this is a "1" circuit. Gotta fix that.

7.6.4 Scaling

I said when I started using the "1" circuit that I would get the right results and then *scale* the circuit elements to get the results where I want them, both in magnitude and in frequency.

Let's start with frequency scaling. In general, frequency scaling means to change the range of the frequency response *without altering the impedance levels*. That's the clue to the transformations. Suppose we want to increase the frequency response range by a frequency-scaling factor k_f.

The impedance of an inductor is $j\omega L$. If I want to increase ω by multiplying it by k_f without changing the impedance, I must *divide* L by k_f. That will make the impedance of the inductor have the same value, but at a frequency k_f times as large.

Here are the frequency-scaling calculations for R, L, and C:

$$Z_R(\omega) = R, \text{ so } k_f\omega \text{ implies } R \text{ unchanged};$$
$$Z_L(\omega) = j\omega L, \text{ so } k_f\omega \text{ implies } L \to L/k_f;$$
$$Z_C(\omega) = 1/j\omega C, \text{ so } k_f\omega \text{ implies } C \to C/k_f.$$

In our case, the "1" circuit has a resonant frequency ω_o of 1 rad/s. We want f_o to be 10 kHz, which says that ω_o is to be $2\pi \times 10000$. Hence

$$k_f = 2\pi \times 10000/1 = 2 \times 10^4\pi,$$
$$L = 1/2 \times 10^4\pi = 15.92 \text{ μH},$$
$$C = 1/2 \times 10^4\pi = 15.92 \text{ μF},$$
$$R = 1\,\Omega, \text{ and } R_L = 0.1\,\Omega.$$

Now the notch should be at 10 kHz. But the input resistance is wrong. Recall in the specifications that we were getting our input from a 600-Ω source. R in our circuit represents the 600-Ω source resistance.

We need magnitude scaling. In general, magnitude scaling means to change the magnitude of the impedance *without altering the frequency response*. That's the clue to the transformations. Suppose we want to increase the magnitude by a magnitude-scaling factor k_m.

The impedance of an inductor is $j\omega L$. If I want to multiply magnitude of that impedance by k^m without changing the frequency, I must multiply L by k_m. Here are the magnitude-scaling calculations for R, L, and C:

$$Z_R(\omega) = R, \text{ so } k_m Z_R \text{ implies } R \to k_m R;$$
$$Z_L(\omega) = j\omega L, \text{ so } k_m Z_L \text{ implies } L \to k_m L;$$
$$Z_C(\omega) = 1/j\omega C, \text{ so } k_m Z_C \text{ implies } C \to C/k_m.$$

In our "1" circuit, $R = 1\,\Omega$. We want this to be 600 Ω, so

$$k_m = 600/1 = 600,$$
$$R = 600 \times 1 = 600\,\Omega,$$
$$L = 600 \times 15.92 \times 10^{-6} = 9.55\,\text{mH},$$
$$C = 15.92 \times 10^{-6}/600 = 0.0265\,\mu\text{F},$$
$$R_L = 600 \times 0.1 = 60\,\Omega.$$

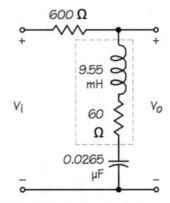

7.6.5 Final Circuit for Whistle Filter

That scaling has given us the final values of the circuit elements. The result is shown in Fig. 7.36 and the frequency response is shown in Fig. 7.37. I haven't tried to reduce this to commercial values.

FIGURE 7.36: Whistle filter (including source R_{Th}).

Does this do it? Here are some check points on the curve:

f (kHz)	IH(f) dB
0.02	0.0
1.0	0.0
5.0	−1.8
9.5	−17.7
10.0	−20.8
10.5	−18.0
20.0	−1.9

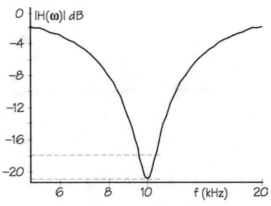

FIGURE 7.37: Frequency response of Whistle filter.

Does this meet the specifications? 0 dB in the passband. Flat to within 3 dB in the passband to 5 kHz. Notch depth of slightly over 20 dB. Bandwidth of close to 1 kHz between the half-power points. Yes.

7.7   SUMMARY

This chapter in a way closes the loop, because one of the important topics in this chapter is the sinusoidal steady state and its relationship to functions in the frequency domain.

Resonance is a specialized part of what happens in general in the frequency domain. Resonance has built around it several general concepts, such as Q and bandwidth. These show up in filter design, for example.

The next chapter, the last one in *Pragmatic Circuits II,* heads off in a somewhat different direction—AC power. This topic is part of the whole area of the sinusoidal steady state, but it's a specialized one with its own terminology.

CHAPTER 8

Power in AC Circuits: SPQR Without a Roma

Power. You already know about power as the product of voltage and current. Power is still voltage times current when we work in the sinusoidal steady state. But there are some new things of interest.

If both voltage and current are sinusoids of the same frequency, their product is the product of sinusoids. Interesting things happen when these sinusoids differ in phase. That's what leads to the various flavors of "power" in AC circuits.

Keep in mind in this chapter that we are dealing solely with circuits operating in the sinusoidal steady state. We are not going to be interested in the natural responses, which in real systems must always die out anyway. We also will be dealing with excitations at a single frequency, so all of our sinusoids will have the same *omega* in their arguments. (If we have excitations at different frequencies, we can treat them separately and then apply superposition, but only for voltage and current. Power isn't linear!)

Before we start into power, though, let's review some circuits in the sinusoidal steady state and the applications of phasors to them. It's been some time since we last worked with those things.

8.1 PHASOR DOMAIN

Phasors represent everything about sinusoidal signals in the steady state except the frequency. That frequency must be the same throughout the whole network. But phasors follow very naturally from what happens in the time domain. So this review of phasors (and of some of the circuit analysis techniques of the dim past) will start with the time domain.

8.1.1 Time-Domain Solution

Consider the circuit shown in Fig. 8.1. The circuit is excited by a sine wave. The frequency is 500 hertz (Hz). Remember that the number before t in a sinusoid must be in radians per second. Also remember that most of us speak in *hertz*, so we can easily mix up the two. There's a 2π needed lots of times:

$$\omega \text{ (radians/second)} = 2\pi f \quad (f \text{ in hertz})$$

With that reminder, let's write time-domain equations for Example I. I am going to write them using *state equations*. Recall that state equations always involve the "energy-storage" variables. So in the circuit of Fig. 8.1 I will use the current through the inductor (flowing to the right at the top) and the voltage across the capacitor (positive at the top).

Here are the two equations, one a mesh equation for the mesh on the left, the other a node equation at the top right:

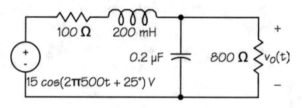

FIGURE 8.1: Example I in the time domain.

$$-15\cos(2\pi 500t + 25°) + 100i_1 + 0.2\frac{di_1}{dt} + v_o = 0,$$

$$-i_1 + 0.2 \times 10^{-6}\frac{dv_o}{dt} + \frac{v_o}{800} = 0,$$

$$v_0(0) = 0, \ i_1(0) = 0.$$

Maple handles these to yield a complete solution:

$$v_o(t) = \left[12.22\cos 2\pi 500t + 5.51\sin 2\pi 500t \right.$$
$$\left. -e^{-3375t}\left(12.22\cos 2\pi 651t + 14.31\sin 2\pi 651t\right)\right]u(t)\,\text{V}$$

But we said we wanted only the steady-state result, so we simply discard those terms that die out as time goes on. The steady-state result is

$$v_{oss}(t) = 12.22\cos 2\pi 500t + 5.51\sin 2\pi 500t$$
$$= 13.40\cos(2000t - 24.3°)\,\text{V}.$$

O.K., fine, nice, etc. Well, the steady-state result certainly wasn't hard to get. We put 15 V in at 500 Hz, we get 13.40 V out, certainly at the same frequency, with a phase angle that says the output lags the input by $25 - (-24.3) = 49.3°$. (Remember that *lags* means *happens later*.)

Let's change the game just a little by using Euler to represent the input in complex-exponential form (j is the square root of -1 instead of i):

$$v_s(t) = 15e^{j(2\pi 500t + 25°)}\,\text{V}.$$

If we put this exponential form (rather than the cosine form) into the differential equations, the solution is still easy to get, although it is messier:

$$v_o(t) = \big[(12.22 - j5.51)e^{j2\pi 500t}$$
$$-e^{-3375t}(12.21 - j5.51)\cos 2\pi 651t$$
$$-e^{-3375t}(14.31 + j4.84)\sin 2\pi 651t\big]u(t)\,\mathrm{V}.$$

If we extract the steady-state portion of the solution by tossing out the terms that die out, we are left with

$$v_{oss}(t) = (12.22 - j5.51)e^{j2\pi 500t}$$
$$= 13.40e^{j(2\pi 500t - 24.3°)}\,\mathrm{V}.$$

Hmmm, that looks right, but is this the same as the original time-domain solution we got about five paragraphs ago? If we apply Euler's formula, we'll get back to a familiar form:

$$v_{oss}(t) = 13.40\cos(2\pi 500t - 24.3°)$$
$$+ j13.40\sin(2\pi 500t - 24.3°)\,\mathrm{V}.$$

Ooops, that wasn't what we had before. If we look at the altered source, though, we can see why. When we changed to a complex exponential, we brought in another term:

$$v_s(t) = 15e^{j(2000t + 25°)}$$
$$= 15\cos(2000t + 25°) + j15\sin(2000t + 25°)\,\mathrm{V}.$$

Writing the time-domain source in this form has added an extra term. But isn't this *superposition*? Isn't our system linear, so that the output due to two sources added together is the same as the sum of the outputs due to each source acting alone? Sure!

This superposition is slightly different from some that we did a few chapters ago, though. The added term has a "flag" attached to it so that we can see what happens to its output later. This "j" flag marks the terms of the output that arise because of the marked term of the input. This tells us that, to select the proper output terms, discard those that are marked with this flag:

$$v_{oss}(t) = 13.40\cos(2\pi 500t - 24.3°)\,\mathrm{V}.$$

That brings us back to the correct time-domain output. (You should recognize removing this "flag" term as "taking the real part.")

8.1.2 Phasor-Domain Solution

Some time ago, we learned to convert circuits into the phasor domain and solve them using complex algebra. Let's do that with Example I using

$$Z_R = R, \quad Z_L = j\omega L, \quad Z_C = 1/j\omega C.$$

The converted circuit is shown in Fig. 8.2.

One node equation will do the job, with the reference node at the bottom and V_o at the top right:

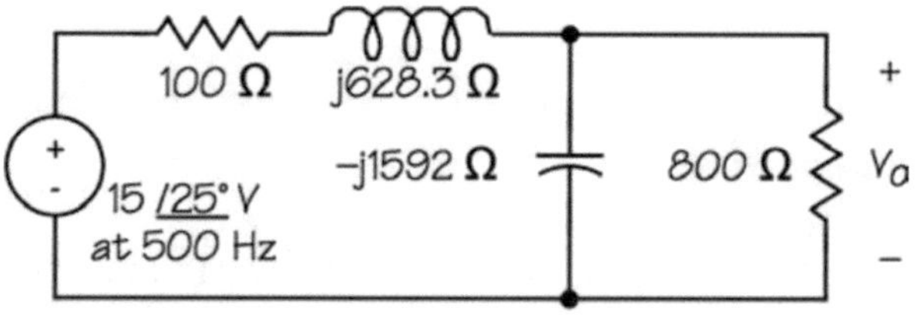

$$\frac{V_o - 15e^{j25°}}{100 + j628} + \frac{V_o}{-j1592} + \frac{V_o}{800} = 0.$$

FIGURE 8.2: Example I in the phasor domain.

Solving this yields

$$V_o = 12.22 - j5.51 = 13.40e^{-j24.3°} \text{ V},$$

which we often abbreviate as

$$V_o = 13.40\angle - 24.3° \text{ V},$$

which converts back to the time domain as

$$v_{oss}(t) = 13.40e^{j(2\pi 500t - 24.3°)}$$
$$= 13.40[\cos(2\pi 500t - 24.3°)$$
$$+ j\sin(2\pi 500t - 24.3°)]\text{V},$$

which when we drop the "flagged" terms becomes

$$v_{oss}(t) = 13.40\cos(2\pi 500t - 24.3°)\text{V}.$$

There! We got the same answer as in the previous section.

8.1.3 Proportionality

Our systems are linear, so we can use techniques such as proportionality to find solutions. Here's Example I to be solved again, this time by "guessing" at the solution and working back to the left to find out how well we did. Fig. 8.3 shows Example I marked for this method.

I'll start by guessing that the output is 1 V at an angle of zero degrees. The rest follows by simple Ohm's-law and Kirchhoff's-laws calculations:

$$V_{o-guess} = 1,$$
$$I_{800} = V_{o-guess}/800 = 0.00125,$$
$$I_C = V_{o-guess}/-j1591 = j0.0006283,$$
$$I_{top} = I_{800} + I_C = 0.00125 + j0.0006283,$$
$$V_{top} = I_{top}(100 + j628) = -0.2698 + j0.8482,$$
$$V_{s-guess} = V_{top} + V_{o-guess} = 0.7302 + j0.8482$$
$$= 1.1192\angle49.3°\,\text{V}.$$

The resultant guess makes the source too small, so we need to boost everything by 15/1.1192. Its phase angle is too large by 49.3° − 25°, so we need to shift all phase angles backward by that amount. The result is the same as we got before:

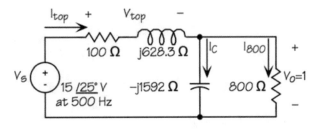

FIGURE 8.3: Example I by proportionality.

$$V_o = 13.40\angle-24.3°\,\text{V}.$$

8.1.4 Coupled Coils

Here's a review example that includes both coupled coils and phasors. Example II is shown in Fig. 8.4 in the time domain.

Note that the frequency is 1.5 MHz, which we must remember to convert to radians per second via 2π before we use it. Writing all the elements as impedances in the phasor domain, we get

$$Z_{1000} = 1000\Omega,\; Z_{47} = -j2257.5\Omega,$$
$$Z_{200} = j1885.0\Omega,\; Z_{60} = j565.5\Omega,\; Z_{50} = j471.2\Omega,$$
$$Z_{220} = -j482.3\Omega,\; Z_{5000} = 5000\Omega.$$

Writing equations for circuits involving coupled coils generally is easier if we use mesh equations. Recall that a current entering the dot on one side of a pair of coupled coils induces a voltage that is positive at the dot on the other side.

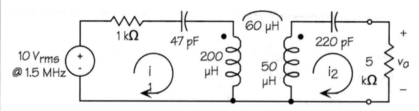

FIGURE 8.4: Example II: phasors and coupling.

$$-10 = (Z_{1000} + Z_{47})I_1 + Z_{200}I_1 - Z_{60}I_2,$$
$$0 = Z_{50}I_2 - Z_{60}I_1 + Z_{220}I_2 + Z_{5000}I_2,$$
$$V_o = Z_{5000}I_2.$$

which yields

$$V_o = -1.6678 + j4.731$$
$$= 5.0165e^{j109.4°} = 5.0165\angle109.4° \, V_{rms},$$

to be translated to the time domain into

$$v_{oss}(t) = 5.0165\cos(2\pi1.5 \times 10^6 t + 109.4°) \, V_{rms}.$$

8.1.5 Thévenin Equivalent

What is the Thévenin equivalent of the circuit of Example II as seen from the 5-kΩ resistor? We will need any two of the open-circuit voltage, the short-circuit current, and the Thévenin-equivalent impedance. The first two are rather easily found from modifications of the original mesh equations.

For the open-circuit voltage, the current I_2 will be zero, so there is only one mesh equation. The output voltage is the voltage induced on the right by the current I_1 on the left.

$$-10 = (Z_{1000} + Z_{47})I_1 + Z_{200}I_1,$$
$$V_{o-oc} = Z_{60}I_1.$$

Solving,

$$V_{o-oc} = 5.299\angle110.4° \, V.$$

The short-circuit current is found from the original equations by replacing the 5-kΩ resistor with zero and solving for I_2:

$$-10 = (Z_{1000} + Z_{47})I_1 + Z_{200}I_1 - Z_{60}I_2,$$
$$0 = Z_{50}I_2 - Z_{60}I_1 + Z_{220}I_2,$$

which gives

$$I_2 = I_{sc} = -0.6265 + j17.89 \, \text{mA}.$$

The Thévenin-equivalent impedance is the ratio of the open-circuit voltage to the short-circuit current:

$$Z_{Th} = V_{o-oc}/I_{sc} = 280.8 + j93.57 \, \Omega.$$

It is easy to verify that this circuit, shown in Fig. 8.5, gives the same result for V_o as did the original circuit. (Don't forget that the results are rms values of voltage and current.)

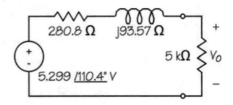

FIGURE 8.5: Example III: Thévenin equivalent.

8.2   AC POWER

AC power has a number of facets, but they all follow from the power relationship

$$p(t) = v(t)i(t).$$

With AC circuits we generally want the steady-state power, so it is easiest to work entirely in the phasor domain to find these results.

8.2.1   Simple Example of AC Power

Let's start with the simple circuit shown in Fig. 8.6.

I want all the steady-state voltages and currents in the time domain, so I'll work the problem in the phasor domain. The converted Example IV is shown in Fig. 8.7.

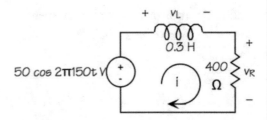

FIGURE 8.6: Example IV: power in the time domain.

Here are all the results:

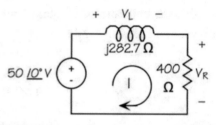

$$I = \frac{50\angle 0}{j282.7 + 400} = 0.1021\angle -35.3° \text{A},$$

$$i(t) = 0.1021\cos(2\pi150t - 35.3°)\text{A},$$

$$V_R = 400I = 40.83\angle -35.3°\text{V},$$

$$v_R(t) = 40.83\cos(2\pi150t - 35.3°)\text{V},$$

$$V_L = j282.7I = 28.86\angle 54.7°\text{V},$$

$$v_L(t) = 28.86\cos(2\pi150t + 54.7°)\text{V}.$$

FIGURE 8.7: Example IV: power in the phasor domain.

Let's start with the power delivered to the resistor. We know that in the time domain this will be the product of the resistor's voltage and current:

$$P_R(t) = v_R(t)i(t)$$
$$= 40.83\cos(2\pi150t - 35.3°)$$
$$\times 0.1021\cos(2\pi150t - 35.3°)$$
$$= 4.168\cos^2(2\pi150t - 35.3°)\text{W},$$

$$\text{but } \cos^2 x = \frac{1}{2}(1 + \cos 2x) \text{ so}$$

$$P_R(t) = 2.084 + 2.084\cos(2\pi300t - 70.5°)\text{W}.$$

Note two things. First, there is a constant term (2.084) that doesn't depend on frequency. Second, there's a cosine term that has twice the frequency of the source. A plot of this power is interesting and is shown in Fig. 8.8.

By integrating this power over one period, we can get the average value of the power delivered to the resistor (although by inspection the answer is going to be 2.084 W):

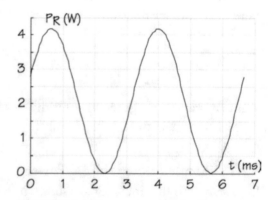

FIGURE 8.8: Power absorbed by the resistor.

$$P_{AV-R} = \frac{1}{T}\int_0^T P_R(t)dt, \text{ where } T = \frac{1}{f} = \frac{1}{150},$$

$$= 2.084\text{W}.$$

So the *average power* delivered to the 400-Ω resistor is 2.084 W, but the power varies with time at *twice the frequency* of the input.

Now consider the power delivered to the inductor. Using some common trig relationships, I get

$$P_L(t) = v_L(t)i(t)$$
$$= 28.86\cos(2\pi150t + 54.7°)$$
$$\times 0.1021\cos(2\pi150t - 35.3°)$$
$$= 28.86\cos(2\pi150t + 54.7°)$$
$$\times 0.1021\sin(2\pi150t + 54.7°)$$
$$= 2.946 \times \frac{1}{2}\sin(2\pi300t + 109.4°)$$
$$= 1.473\cos(2\pi300t + 19.4°)W.$$

Let's plot this result in Fig. 8.9 and see what it looks like.

Note that it is pretty obvious that the average value of this function is zero. But there is still power being delivered to the inductor, albeit half the time the power is being delivered and half the time it is being removed. So energy is, in effect, sloshing back and forth, into and out of the inductor. Furthermore, there are two round trips during each cycle of the original input.

Next consider the power dselivered from the source to the circuit:

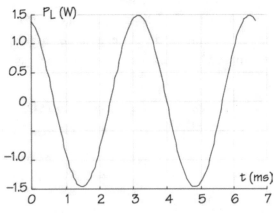

FIGURE 8.9: Power absorbed by the inductor.

$$P_s(t) = v_s(t)i(t)$$
$$= 50\cos 2\pi150t \times 0.1021\cos(2\pi150t - 35.3°)$$
$$= 5.104\cos 2\pi150t \times$$
$$\left[\cos 2\pi150t \cos(-35.3°)\right.$$
$$\left. -\sin 2\pi150t \sin(-35.3°)\right]$$
$$= 5.104\cos^2 2\pi150t \cos(-35.3°)$$
$$- 5.104\cos 2\pi150t \sin 2\pi150t \sin(-35.3°)$$

$$= 2.552(1 + \cos 2\pi 300t)\cos(-35.3°)$$
$$- 2.552 \sin 2\pi 300t \sin(-35.3°)$$
$$= 2.552 \cos(-35.3°) + 2.552 \cos(2\pi 300t - 35.3°)$$
$$= 2.084 + 2.552 \cos(2\pi 300t - 35.3°)\,\text{W}.$$

The first term is the average power delivered to the resistor. The second contains the cosine part of the power to the resistor and the "sloshing" for the inductor). Fig. 8.10 is a plot of the power delivered by the source.

The plot shows that power is delivered, on the average, to the circuit. But note that there are intervals of time when power is actually returned to the source.

As a check, let's compute the source power by summing the power delivered to the resistor and to the inductor:

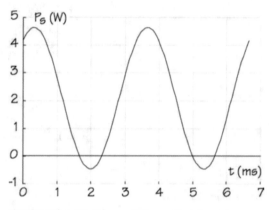

FIGURE 8.10: Power delivered by the source.

$$P_s(t) = P_R(t) + P_L(t)$$
$$= \left[2.084 + 2.084 \cos(2\pi 300t - 70.5°)\right]$$
$$+ 1.473 \cos(2\pi 300t + 19.4°)$$
$$= 2.084 + 2.552 \cos(2\pi 300t - 35.2°)\,\text{W}.$$

Great! This comes out the same, which of course it had to do. But it is nice to know that I've done things correctly, or at least made the same mistakes in both results.

8.2.2 General Case

If we are going to see how all this can be done in the phasor domain, we need to look at a simple power problem in general terms rather than with numbers. Consider the circuit element shown in Fig. 8.11.

Let's define the voltage and the current as sinusoids with phase angles:

$$v(t) = V_p \cos(2\pi ft + \theta_1),$$
$$i(t) = I_p \cos(2\pi ft + \theta_2).$$

Since phase angles define starting points, we won't lose anything by making the phase of the current zero. In effect, this makes the current the *reference phase*, or the phase that we consider to be the starting point of our time axis. (Besides, it simplifies the trig tremendously!) So our voltage and current become

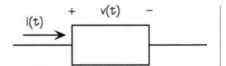

FIGURE 8.11: General case.

$$v(t) = V_p \cos(2\pi ft + \theta),$$
$$i(t) = I_p \cos 2\pi ft.$$

The following calculations are mostly messy trig. A useful relationship is

$$\cos x \cos y = \frac{1}{2}\cos(x + y) + \frac{1}{2}\cos(x - y).$$

Using that and skipping some of the details, I get

$$p(t) = v(t)i(t) = V_p \cos(2\pi ft + \theta) \times I_p \cos 2\pi ft$$
$$= \frac{V_p I_p}{2}\left[\cos(2\pi ft + \theta + 2\pi ft) + \cos(2\pi ft + \theta - 2\pi ft)\right]$$
$$= \frac{V_p I_p}{2}\cos\theta + \frac{V_p I_p}{2}\cos 4\pi ft \cos\theta$$
$$- \frac{V_p I_p}{2}\sin 4\pi ft \sin\theta$$
$$= \frac{V_p I_p}{2}(1 + \cos 4\pi ft)\cos\theta - \frac{V_p I_p}{2}\sin 4\pi ft \sin\theta.$$

Well, how about that! There are lots of things to note:

- Both terms contain double frequency components, $2 \times 2\pi f$, so power in AC systems wiggles at twice the frequency of the input.
- The first term has an average value, which is

$$P_{AV} = \frac{V_p I_p}{2}\cos\theta,$$

and this term represents net transfer of energy into the circuit.

- The first term also has a varying component, so energy is not transferred at a constant level.
- The second term has a zero average value, so it does not represent any net transfer of energy into the circuit.
- The second term represents energy that comes and goes, power that is delivered during one *quarter* of the input cycle, then removed during the next quarter, and so on.

All of this enables us to define several basic terms of AC power:

Average power is the power that, on the average over one cycle of the input, is delivered to the circuit. Average power is

$$P = \frac{V_p I_p}{2} \cos\theta,$$

which has the unit of *watts* with the symbol W. The phase angle θ is the phase angle of the voltage if the angle of the current is zero. (It's the angle of the voltage minus the angle of the current in the general case.)

Average power is often called *real power* or *active power* or sometimes, sloppily, *power*. The common symbol is P.

Reactive power represents energy that sloshes back and forth, into and out of the circuit. Reactive power is

$$Q = \frac{V_p I_p}{2} \sin\theta.$$

Units are interesting. Clearly this is voltage times current so the unit should be *watts*. But that confuses things, so instead we always use *volt-amperes reactive*. This is abbreviated VAR and is generally pronounced as a word.

The common symbol for reactive power is Q.

If the angle is positive, Q is positive, which is the situation for an inductor. For a capacitor, Q is negative.

Power factor is a measure of how much the phase angles of the voltage and the current differ. Power factor is

$$pf = \cos\theta.$$

Since the cosine is always positive for angles between $-90°$ and $+90°$, we can't tell whether the voltage angle relative to the current's angle is larger or smaller. This turns out to be important, especially considering that the sine of the angle governs the sign of the reactive power Q.

If $\theta > 0$, then the voltage has a larger phase angle than the current. This means that, on the time axis, the voltage "happens first." So we might say that the voltage *leads* the current.

But don't! Get in the habit of always saying the word *current* first in such statements. Hence if $\theta > 0$, we say *the current lags the voltage.* That means the circuit must be inductive (because there must be some voltage acting on the inductor before current is forced to flow[1]).

Stating a power factor without adding either *lagging* or *leading* is making an incomplete statement. Inductive circuits are *lagging* and their reactive power Q is *positive*. The opposite is true for capacitive circuits.

Recall Example IV where we computed voltages, currents, and powers both in the time and the phasor domains. The results in the phasor domain are shown in Fig. 8.12.

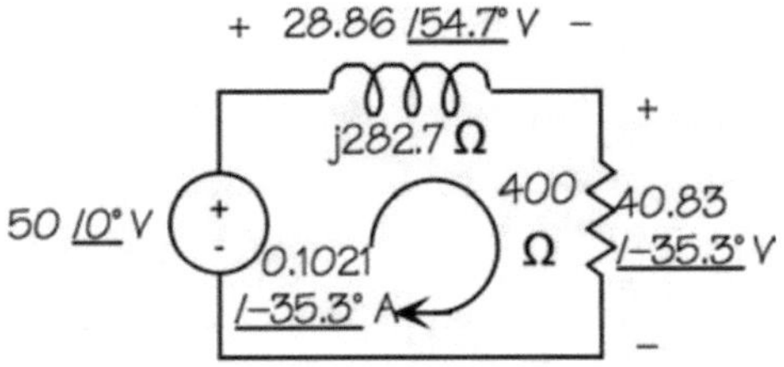

Using the voltages and currents shown in Fig. 8.12, let's compute P, Q, and the power factor for each element.

FIGURE 8.12: Results of Fig. 8.7 .

For the resistor,

$$P_R = \frac{40.83 \times 0.1021}{2}\cos(-35.3° - (-35.3°)) = 2.084\,\text{W},$$

$$Q_R = \frac{40.83 \times 0.1021}{2}\sin(-35.3° - (-35.3°)) = 0,$$

$$pf_R = \cos(-35.3° - (-35.3°)) = 1.$$

This power factor is usually spoken of as being *unity* power factor.

For the inductor,

$$P_L = \frac{28.86 \times 0.1021}{2}\cos(54.7° - (-35.3°)) = 0,$$

$$Q_L = \frac{28.86 \times 0.1021}{2}\sin(54.7° - (-35.3°)) = 1.473\,\text{VAR},$$

$$pf_L = \cos(54.7° - (-35.3°)) = 0 \text{ lagging.}$$

Note the use of the word *lagging* here. The element is inductive. (Sure! It's an inductor!)

For the overall circuit,

---

[1]This is a sloppy approach to inductors but it gives a simple way of remembering what must happen first, i.e., current in an inductor lags the voltage across it.

$$P_s = \frac{50 \times 0.1021}{2} \cos(0° - (-35.3°)) = 2.0\,\text{W},$$

$$Q_s = \frac{50 \times 0.1021}{2} \sin(0° - (-35.3°)) = 1.473\,\text{VAR},$$

$$pf_s = \cos(0° - (-35.3°)) = 0.817 \text{ lagging.}$$

Here too, the overall circuit is inductive. When we say that, we mean that the power factor is lagging and that the reactive power Q is positive.

8.2.3 Complex Power

Let's go back to our general case, repeated here in Fig. 8.13. This time, though, instead of using peak values for the amplitudes of the cosine terms, let's use *rms*. After all, we generally speak in *rms* and our meters indicate *rms*.

FIGURE 8.13: General case.

For current and voltage I'll not only use *rms* but I will also include a phase angle for each of them:

$$v(t) = V_{rms} \cos(2\pi ft + \theta_1),$$
$$i(t) = I_{rms} \cos(2\pi ft + \theta_2).$$

Now shift to the phasor domain in exponential form:

$$V = V_{rms} e^{j\theta_1},$$
$$I = I_{rms} e^{j\theta_2}.$$

I am going to do something which at first will seem strange—I am going to define *complex power* (symbol S) as the product of voltage and current in the phasor domain. But this comes with a twist: I'll use the *complex conjugate* of the current. (Remember that the complex conjugate is the original expression with every j replaced by $-j$.)

Here's S for our general example:

$$S = V \times I^*$$
$$= V_{rms} e^{j\theta_1} \times \left(I_{rms} e^{j\theta_2}\right)^* = V_{rms} e^{j\theta_1} \times I_{rms} e^{-j\theta_2}$$
$$= V_{rms} I_{rms} e^{j(\theta_1 - \theta_2)}.$$

Now apply Euler's formula to get

$$S = V_{rms}I_{rms}\cos(\theta_1 - \theta_2) + jV_{rms}I_{rms}\sin(\theta_1 - \theta_2).$$

Whoa! The first term is just the average power P, this time written in rms and with the difference of the voltage and current phase angles explicitly shown. The second term is the reactive power, written in the same way. (The $1/2$ is now missing because we are using rms.) Aha! That says

$$S = P + jQ.$$

This S is called *complex power* and is, of course, the product of voltage and current. But to distinguish it from average power, instead of watts we use *volt-amperes* and a symbol VA.

There are several things we can do with this.

First, we define *apparent power* as simply the product of the amplitudes of the voltage and the current:

$$|S| = V_{rms}I_{rms}.$$

The units of apparent power are also *volt-amperes* or VA. Apparent power is useful both because it is the magnitude of the complex power and because it is easily gotten from simple meter readings of voltage and current. The phase angle is being disregarded.

The power factor is the cosine of the angle between the voltage and the current, but we can get it in another form, again one that is easy to measure:

$$pf = \cos(\theta_1 - \theta_2) = \frac{V_{rms}I_{rms}\cos(\theta_1 - \theta_2)}{V_{rms}I_{rms}} = \frac{P}{|S|}$$

$$= \frac{\text{average power}}{\text{apparent power}}.$$

Another useful fact about complex power is that, since it is complex, we can draw triangles. (Oh, gee, really? Really draw triangles? Oooo!) Fig. 8.14 gives the complex relationship of S, P, and Q. A more common form of the same information is shown as the *power triangle* in Fig. 8.15.

We will use this graphical approach in an example later.

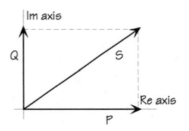

FIGURE 8.14: Relationship of P, Q, and S.

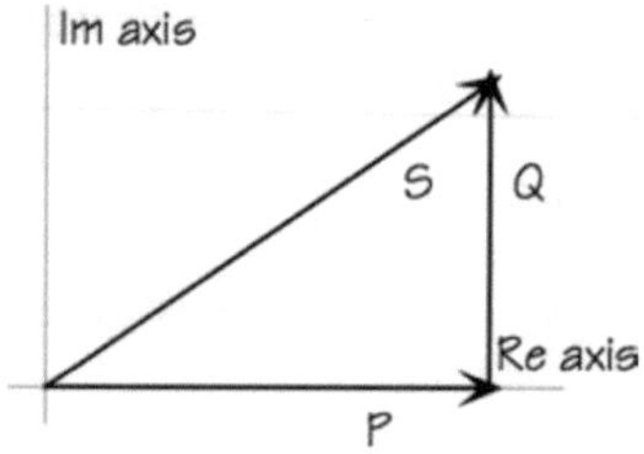

Keep in mind that, while these look like phasor diagrams, S, P, and Q are *not* phasors, just as complex impedances are not phasors. (Even if it quacks like a duck….)

We know that energy is conserved and also that power is:

- Since average power P is the actual power being delivered, P must be conserved.
- Reactive power Q is merely energy that is swapped back and forth. The "backs" must balance the "forths," so there is no net energy flow. Hence Q is conserved.
- Complex power S is the complex sum of P and Q, so S must also be conserved.

Now it is time for some examples.

FIGURE 8.15: Power triangle.

8.3   EXAMPLES

AC power has lots of new concepts that didn't arise when you did simple power calculations in DC circuits. Not only do we have average power, but we must now deal with reactive power, complex power, apparent power, and power factor. Our simple unit of watts has been supplemented by VARs and VAs. Power factor requires the addition of *leading* or *lagging*, and we must remember to say *current* first.

These concepts are easier to understand through several examples. I'll try to keep them pretty straightforward; power problems can get hairy very fast.

8.3.1   Example V

The circuit of Fig. 8.16 is a simple model of a possible source, with its own losses (1 Ω), and a capacitive load. The goal is to find the average and the reactive power for each element, the average, reactive, and complex power for the source, the power factor at the source, and the efficiency of the energy delivery to the load.

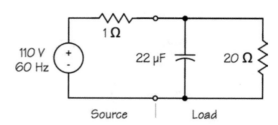

FIGURE 8.16: Example V in the time domain.

I've designated the source and the load on the drawing. The final question, what is the efficiency, takes into account that some energy is lost in the 1-Ω resistance associated with the source.

We start by transforming the circuit into the phasor domain as shown in Fig. 8.17.

Note how the source voltage is stated. Is this rms or peak? When we deal with AC circuits in the steady state, it is a very good bet that the value is rms. That's the only common number that engineers use.

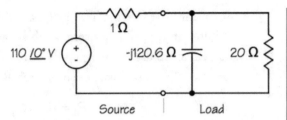

FIGURE 8.17: Example V in the phasor domain.

The time-domain circuit gives nothing about the phase angle of the source. This too is common. It means we are at liberty to choose the phase reference anywhere we want. I have chosen to give the source the zero phase angle; in many problems there are other voltages or currents that can be a useful reference while simplifying the problem.

Finally, the frequency is given in hertz. That's what almost everyone means by "frequency;" radians per second are for mathematics.

First let's find the impedance of the load. The two elements in parallel give us

$$Z_C = 1/(j2\pi60 \times 22 \times 10^{-6}) = -j120.57\,\Omega,$$
$$Z_{load} = \frac{20Z_c}{20 + Z_C} = 19.465 - j3.229\,\Omega.$$

Now for the current through the whole circuit and the voltage across the load:

$$I_s = \frac{110\angle0°}{1 + Z_{load}} = 5.245 + j0.8275\,\text{A},$$
$$V_{load} = Z_{load}I_s = 104.76 - j0.8275$$
$$= 104.76\angle - 0.45°\,\text{V}.$$

From this we can get the current through each of the elements of the load:

$$I_C = V_{load}/Z_C = 0.00686 + j0.8688\,\text{A},$$
$$I_R = V_{load}/20 = 5.238 - j0.0414\,\text{A}.$$

The average and reactive power values for each element can be calculated in several ways. One would be to use V_{load} and the conjugate of the current through the element. You might want to try this.

Another way is to use our heads and some knowledge that we really already have:

- Consider the resistor. The magnitude of the voltage across it is 104.76 V, the magnitude of V_{load}. The power absorbed by the resistor is still v-squared over resistance, provided we use the magnitude of the AC voltage in rms. The reactive power for the resistor is zero because resistors don't absorb reactive power.

$$P_R = 104.76^2/20 = 548.7\,\text{W},$$
$$Q_R = 0.$$

- Similarly, the reactive power absorbed by the capacitor is v-squared over the magnitude of the capacitor's impedance, again using the magnitude of the AC voltage in rms. But here we must also remember to insert a minus sign because the reactive power to a capacitor is negative. The average power will be 0.

$$P_C = 0,$$
$$Q_C = -104.76^2/120.57 = -91.02\,\text{VAR}.$$

The power delivered to the 1-Ω resistance of the source can also be gotten by thinking a little. The current through that resistor is known. The average power delivered to it will therefore be i-squared times the resistance, provided we use the magnitude of the current in rms:

$$I_s = I_C + I_R$$
$$= 0.00686 + j0.8688 + 5.238 - j0.0414$$
$$= 5.245 + j0.8274 = 5.310\angle 9.0°\,\text{A}.$$
$$P_1 = 5.310^2 \times 1 = 28.20\,\text{W},$$
$$Q_1 = 0.$$

Let's compute the source numbers by finding the complex power. We know both the current and the voltage:

$$S_s = 110 \times I_s^* = 110(5.245 - j0.8274)$$
$$= 576.9 - j91.02\,\text{VA},$$
$$P_s = 576.9\,\text{W},$$
$$Q_s = -91.02\,\text{VAR}.$$

These figures can be checked by adding the various numbers for the power absorbed by the circuit elements.

The power factor is the cosine of the angle between the source voltage and the source current. This is the same angle that appears in the complex power. I'll find the power factor that way:

$$pf = \cos\left(\tan^{-1}(Q_s/P_s)\right) = 0.988 \text{ leading.}$$

I appended "leading" because I know the circuit is capacitive. You'll often hear an engineer give a power factor in percent, as "the power factor is 98.8 % leading."

Finally, what is the efficiency? We are concerned only with delivery of useful energy, which means average power. So efficiency will be the ratio of the average power that reaches the load to the average power delivered by the source:

$$efficiency = 100 \times P_R/P_s = 95.11\%.$$

I have saved work in a number of spots by just remembering some of the things I knew from DC circuits.

8.3.2 Example VI

Find P, Q, S, and the power factor of the load Z shown in Fig. 8.18, where $Z = 1.2 + j3.3$ Ω.

After what we did on the previous example, this one is easy. We know the impedance. What we don't is how the circuit is built inside the box. But we don't need to know this. We can mentally model the impedance as a 1.2-Ω resistor in series with an inductor whose reactance is 3.3 Ω.

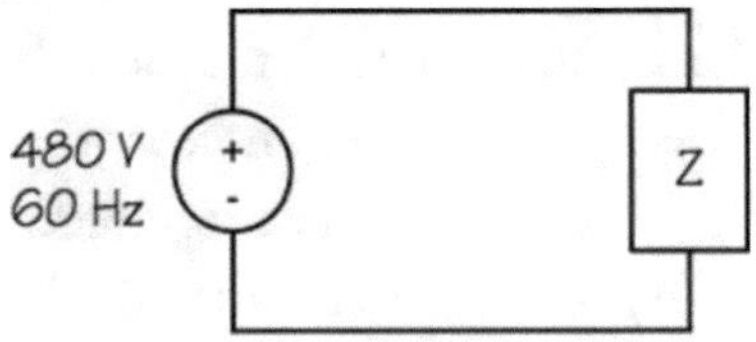

The steps become straightforward. First, find the current through the impedance. Then use the "DC" relationships of i-squared times resistance and i-squared times reactance to get P and Q. S is their complex sum. The power factor is the cosine of the angle of the complex power.

FIGURE 8.18:   Example VI.

$$I = 480/Z = 46.72 - j128.47$$
$$= 136.70\angle -70.0^\circ \text{A},$$
$$P_Z = 136.70^2 \times 1.2 = 22424 \text{W},$$
$$Q_Z = 136.70^2 \times 3.3 = 61667 \text{VAR},$$
$$S_Z = P_Z + jQ_Z = 22424 + j61667 \text{VA},$$
$$pf = \cos\left(\tan^{-1}(61667/22424)\right)$$
$$= 0.342 \text{ lagging.}$$

I made Q positive because I know the impedance is inductive (the imaginary term of Z is positive), and I appended "lagging" to the power factor for the same reason.

We can check our work by computing S for the source (note I conjugate):

$$S_s = 480 \times I^* = 480(46.72 + j128.47)$$
$$= 22426 + j61666 \, \text{VA.}$$

Right.

8.3.3 Example VII

The transformer that supplies a small business is rated at 230 V, 60 Hz, 25 kVA. How large a lighting load can be supplied by this transformer without exceeding its rating? What will be the transformer current?

We are given the transformer's source voltage and the *apparent power* rating. This means that the maximum value of the *magnitude* of S is 25 kVA. Since a lighting load is essentially resistive, at least for incandescent lamps after they are on, the maximum load for lighting is 25 kW. The current will be simply the power divided by the voltage:

$$I = P/V = 25000/230 = 108.7 \, \text{A.}$$

Now consider an induction motor. How large an induction motor can be run from this transformer if, at full load, the motor's power factor is 82% lagging and the motor's efficiency is 88%? Specify the motor in horsepower (1 hp = 746 W).

We know the magnitude of the complex power that the transformer can deliver (25 kVA). The real part of this is the power available to actually run the motor. The power factor is the cosine of the angle of the complex power, which we know. So the average power is the power factor times the magnitude of the complex power:

$$P_{motor} = pf \times S_{mag} = 0.82 \times 25000 = 20500 \, \text{W,}$$
$$hp = efficiency \times P_{motor}/746 = 24.18 \, \text{hp.}$$

The efficiency enters into the second calculation because only 88% of the power delivered to the motor actually is passed through to the shaft. The factor 746 is watts per horsepower.

8.3.4 Example VIII

The circuit of Fig. 8.19 represents three loads connected to a feeder (a pair of wires). These wires each have a line impedance as shown. The loads are

1. an induction motor rated at 230 V, 10 hp, 81% lagging, 90% efficiency,
2. a heater rated at 5 kW at 230 V, and
3. an induction heater rated at 230 V, 10 kVA, 73% lagging.

The line voltage at the load V_L must be 230 V if we are to supply the loads properly. I choose this voltage as my phase-angle reference, so $V_L = 230 \angle 0° \text{ V}$

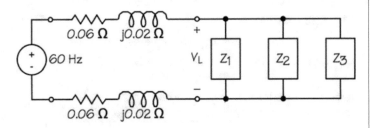

We are to find the current in the line, the power lost in the line, the voltage at

FIGURE 8.19: Example VIII: several loads.

the source, P, Q, S, and pf for the source, and the system's efficiency.

A fairly simple approach to this problem is to find the complex power for each load, then add these to get the complex power for the loads together. From this we can get the line current. The source voltage will be the load voltage plus the losses in the lines. Finally, the power figures for the source will come from adding the complex power for the lines to that of the loads.

As we go along, let's draw the power triangles. They are helpful in visualizing what is taking place.

Load 1 is 10 horsepower, but it's only 90% efficient. So we must convert horsepower into watts and adjust this for the efficiency (*divide by* 0.90 because the motor absorbs more energy than the shaft delivers).

$$P_1 = 746 \times hp/efficiency = 746 \times 10/0.90$$
$$= 8289 \text{ W}.$$

The power triangle of Fig. 8.20 helps us find S. We know that the "P" side of the triangle is the number just found and that the angle is the angle whose cosine is 0.81.

We use some trig to find S:

$$S_1 = P_1 + jP_1 \tan(\cos^{-1} 0.81)$$
$$= 8289 + j6001 \text{ VA}.$$

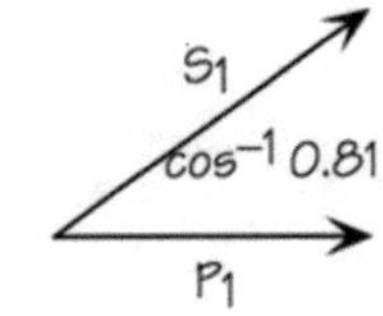

FIGURE 8.20: Load 1: motor.

Load 2 is a heater that is purely resistive. I'll draw the power triangle (Fig. 8.21) just to be complete, but it is trivial.

$$S_2 = 5000 + j0\,\text{VA}.$$

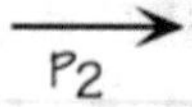

FIGURE 8.21: Load 2: heater.

Load 3 is also inductive. We are given the magnitude of S (kVA) and the power factor. The power triangle of Fig. 8.22 shows this.

Here again, some trig will give us the two sides of the triangle, given that we know the hypotenuse:

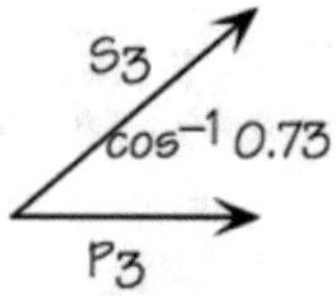

FIGURE 8.22: Load 3: induction heater.

$$\begin{aligned} S_3 &= |S| \times pf + j|S|\sin(\cos^{-1} pf) \\ &= 10000 \times 0.73 + j10{,}000 \times \sin(\cos^{-1} 0.73) \\ &= 7300 + j6834\,\text{VA}. \end{aligned}$$

Now we can add the three values of complex power to get the total for the loads together. Fig. 8.23 is the power triangle for this, drawn roughly to the scale of the first three.

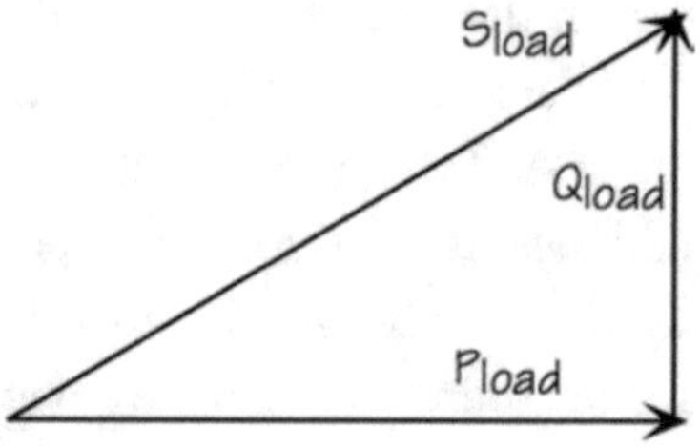

$$S_{load} = S_1 + S_2 + S_3 = 20589 + j12835\,\text{VA}.$$

FIGURE 8.23: total load.

The current flowing into the load can be found from the complex-power relationship, voltage times the conjugate of the current. Knowing this current, we find the power lost in the lines (i-squared times resistance) and the voltage drop in the two wires. (Note the complex conjugate and the sign change for j.)

$$\begin{aligned} I_{line} &= \left(S_{load}/V_L\right)^* = (20589 - j12835)/230 \\ &= 89.52 - j55.80\,\text{A}, \\ P_{line} &= |89.52 - j55.80|^2 (2 \times 0.06) = 1335\,\text{W}, \\ V_{line} &= 2 \times Z_{line} \times I_{line} \\ &= 2(0.06 + j0.02)(89.52 - j55.80) \\ &= 12.974 - j3.115\,\text{V}. \end{aligned}$$

Now add this voltage drop to the load voltage to give the source voltage.

$$\begin{aligned} V_s &= V_L + V_{line} = 230 + 12.97 - j3.115 \\ &= 243.0 - j3.115 = 243.0\angle -0.7°\,\text{V}. \end{aligned}$$

Finally, from the source voltage and the line current we get the complex power for the source, and from that the power factor.

$$S_s = V_s \times I_{line}^* = (243.0 - j3.115)(89.52 + j55.80)$$
$$= 21927 + j13281 \, \text{VA},$$
$$pf = \cos\left(\tan^{-1}\left(\text{Im}[S_s]/\text{Re}[S_s]\right)\right) = 85.5\% \, \text{lagging}.$$

The efficiency is the ratio of the average power delivered to the load (the real part of the load's complex power) to the average power delivered by the source (the real part of its complex power).

$$\textit{efficiency} = 100\frac{\text{Re}[S_{load}]}{\text{Re}[S_s]} = 100\frac{20589}{21927} = 93.9\%.$$

Summarizing all this,

$$I_{line} = 89.52 - j55.80 = 105.49\angle - 31.9° \, \text{A},$$
$$P_{line} = 1335 \, \text{W},$$
$$V_s = 243.0\angle - 0.7° \, \text{V},$$
$$P_s = 21,927 \, \text{W},$$
$$Q_s = 13,281 \, \text{VAR},$$
$$S_s = 21,927 + j13,281 \, \text{VA},$$
$$pf = 85.5\% \, \text{lagging},$$
$$\textit{efficiency} = 93.9\%.$$

8.3.5 Example IX

Add something to the load of Example VIII so that the power factor at the load is raised to 90% lagging. Find the efficiency and the source voltage after this is done.

Fig. 8.24 shows the circuit for this. Why did I choose a capacitor? Because it will absorb negative VARs to help cancel the inductances' positive VARs.

One of the easiest ways to work this problem is through a power triangle. (I find the problem hard to visualize without it.) Fig. 8.25 shows the power triangle for the load.

The power factor of the original load is

$$pf_{load} = \cos\left(\tan^{-1}\left(\text{Im}[S_{load}]/\text{Re}[S_{load}]\right)\right)$$
$$= \cos\left(\tan^{-1}\left(12835/20589\right)\right) = 84.9\% \, \text{lagging}.$$

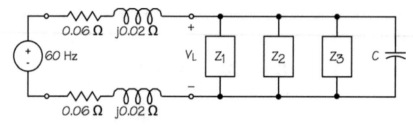

FIGURE 8.24: Example IX: power-factor correction.

To bring this to 90% we have only two things we can change: the average power delivered to the load (the bottom of the triangle) or the reactive power delivered to the load (the right side of the triangle). Since we want the load to do the work we had given it to do, the average power can't change. Hence the reactive power must.

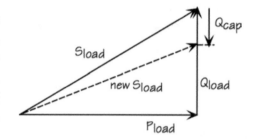

FIGURE 8.25: Power-factor correction.

The power factor is presently too small. Hence the angle on the left of the triangle is too large. Making this angle smaller, thereby raising the power factor, requires that we reduce the reactive power. The amount is the downward arrow on the right in the power triangle.

The new reactive power must satisfy a new power triangle. The base of the triangle is still the same average power to the load. The cosine of the new angle on the left must now be 0.9. The tangent of that angle must be the new reactive power divided by the present average power. Hence

$$\cos\tan^{-1}\left(\frac{Q_{load-new}}{P_{load}}\right) = 0.9,$$

$$Q_{load-new} = P_{load}\tan(\cos^{-1}0.9)$$
$$= 9972\,\text{VAR}.$$

The capacitor must therefore provide some negative VARs:

$$Q_C = Q_{load-new} - Q_{load}$$
$$= 9762 - 12835 = -2863\,\text{VAR}.$$

How much capacitance is this? Recall that we can find Q much as we would in DC circuits. Here it will be the magnitude of the voltage-squared divided by the magnitude of the capacitor's impedance:

$$|Q_c| = \frac{230^2}{|Z_c|} = \frac{230^2}{1/(2\pi 60 C)},$$
$$C = 143.6\,\mu F.$$

(This is quite a bit of capacitance rated at 230 V. Polarized electrolytics can't be used here, either.)

The new load, including the capacitor, is

$$S_{load-new} = 20589 + j9972\,VA.$$

From this, using the same calculation as before, we find

$$efficiency = 94.5\%,$$
$$|V_s| = 242.5.$$

Is it worth the money to install those capacitors? Certainly not from an efficiency standpoint (94.5% versus 93.9%). But if the power company charges us a penalty for having a power factor below 90%, this may make economic sense. The design example in the next section looks at some of the economic tradeoffs.

8.4 POWER(FUL) EXAMPLES
All of the examples here are based on supplying a load through a somewhat lossy line.

8.4.1 Example X
A load of 10 + j7.5 Ω is to be supplied with 480 V through a lossy line, each wire of which has an impedance of 0.2 + j0.5 Ω. The circuit is shown in Fig. 8.26.

Our job is to find the current in the line (*I*), the average power (*P*), the reactive power (*Q*), and the total power (*S*) delivered to the load, and the power factor of the load.

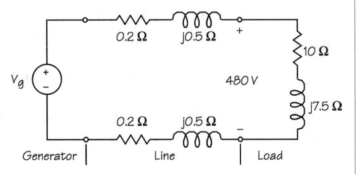

FIGURE 8.26: Examples X and XI.

Let's start by finding the line current:

$$V_{load} = 480 \text{ V},$$
$$Z_{load} = 10 + j7.5 \,\Omega.$$
$$I_{load} = \frac{V_{load}}{Z_{load}} = 30.72 - j23.04 \text{ A}$$
$$= 38.40\angle - 36.9° \text{ A}.$$

From this, I can find the total power S as the load voltage times the complex conjugate of the load current:

$$S_{load} = V_{load} I_{load}^* = 14{,}746 + j11{,}059 \text{ VA}.$$

P and Q are easily picked out of S:

$$P_{load} = \text{Re}\left[S_{load}\right] = 14.7 \text{ kW},$$
$$Q_{load} = \text{Im}\left[S_{load}\right] = 11.1 \text{ kVAR}.$$

There are several ways to compute the power factor, but since I now have P and S, the easy way is

$$pf = \frac{P_{load}}{\left|S_{load}\right|} = 0.80 \text{ lagging.}$$

Don't forget that the power factor must have "units," too, namely lagging or leading.

8.4.2   Example XI

Continue with the example to find the power loss in the line, the generator voltage, and the efficiency of power delivery.

From the line impedance (the sum of the impedances of the two wires), I can get the voltage "lost" in the line:

$$Z_{line} = 2(0.2 + j0.5) = 0.4 + j1.0 \,\Omega,$$
$$V_{line} = Z_{line} I_{load} = 35.33 + j21.50 \text{ V}.$$

The generator voltage is the sum of the load and the line voltages:

$$V_{gen} = V_{line} + V_{load} = 515.33 + j21.50$$
$$= 515.78\angle 2.4° \text{ V.}$$

The power lost in the line can be found by the resistance of the line times the square of the magnitude of the line current:

$$P_{line} = \text{Re}\left[Z_{line}\right]\left|I_{load}\right|^2 = 589.8 \text{ W.}$$

Finally, the efficiency of power delivery is the ratio of the power that reaches the load divided by the power delivered by the generator:

$$eff = 100\frac{P_{load}}{P_{load} + P_{line}} = 96.2 \%.$$

8.4.3 Example XII

A new load is installed in the plant. This load, in parallel with the first load, is somewhat capacitive: $15 - j10$ Ω. The power company readjusts the incoming voltage to maintain 480 V at the load (perhaps by changing taps on the transformer). The new load is shown in Fig. 8.27.

Answer all the same questions as before, but for this new load combination.

The new total load is

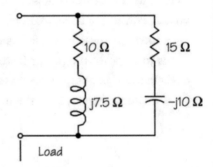

FIGURE 8.27: Example XII and XIII modified load.

$$Z_{add} = 15 - j10,$$
$$Z_{new} = Z_{add} \| Z_{load} = 8.861 + j1.386 \text{ Ω},$$

which makes the load current

$$I_{load} = \frac{V_{load}}{Z_{new}} = 52.87 - j8.271 \text{ A.}$$

Finding the load's S, P, Q, and the power factor as before yields

$$S_{load} = V_{load}I_{load}^* = 25{,}380 + j3970 \text{ VA,}$$

$$P_{load} = \text{Re}\left[S_{load}\right] = 25.4 \text{ kW,}$$

$$Q_{load} = \text{Im}\left[S_{load}\right] = 3.97 \text{ kVAR,}$$

$$pf = \frac{P_{load}}{\left|S_{load}\right|} = 0.988 \text{ lagging.}$$

The voltage across the line impedance and the generator voltage are now

$$V_{line} = Z_{line}I_{load} = 29.42 + j49.57 \text{ V,}$$

$$V_{gen} = V_{line} + V_{load} = 509.4 + j49.57 \text{ V}$$

$$= 511.8\angle 5.6° \text{ V.}$$

There's a surprise here! The new generator voltage is lower than the voltage before adding a second load. This doesn't appear on the surface to make sense. After all, the new load draws more current, so the line should have a greater effect on the generator voltage.

However, the added load is capacitive, which means a leading power factor. The net effect is to improve the power factor of the total load, making it closer to unity and reducing the reactive components of the voltages and currents.

Finishing the job, I get the line loss to be

$$P_{line} = \text{Re}\left[Z_{line}\right]\left|I_{load}\right|^2 = 1146 \text{ W}$$

and the efficiency to be

$$eff = 100\frac{P_{load}}{P_{load} + P_{line}} = 95.7\%.$$

Note that, even with the reduction in generator voltage, the efficiency did not get better. It went down because there is now more current flowing in the line and the loss is greater (approximately double).

8.4.4 Example XIII

Suppose, in the example we just did, that the power company can't be convinced to readjust the generator voltage to keep the load voltage constant at 480 V. Instead, they supply 480 V at the generator. (They are not required to make an adjustment, because their tariffs usually allow them to supply a "nominal voltage" of some value, plus or minus as much as 10%.)

What will the load voltage be after adding the second load?

The total impedance seen by the generator is

$$Z_{total} = Z_{line} + Z_{load} = 9.261 + j2.386 \ \Omega.$$

This makes the current

$$I = \frac{V_g}{Z_{total}} = \frac{480}{9.261 + j2.386} = 48.60 - j12.52 \ \text{A}.$$

The voltage at the load is now

$$V_{load} = Vg - Z_{line} I = 448.0 - j43.6 \ \text{V}$$
$$= 450.2\angle - 5.6° \ \text{V}.$$

While the plant operators might not be happy with a load voltage that's 30 V low, the power company is probably still within its tariffed voltage. At the same time, the lower voltage might mean lower sales of kilowatt-hours, which means lower profit, which means somebody better decide to come out to the plant and readjust the voltage.

How much difference can this make in power sold? Consider that the load voltage is about 6% low. Since power is proportional to the square of the voltage, the power will be about 12% less: $(1 - 0.06)2 \approx 1 - 0.12$. Since the load power is 25.4 kW at the proper voltage of 480 V, the power company is losing the sale of about 3.0 kW each hour. If the plant runs 24 h a day seven days a week, that's over 2000 kWhr per month. If industrial rates are 5¢ per kWhr, that's $100—every month. For that, they could afford to send someone to make the adjustment.

8.5   DESIGN EXAMPLE

It's Tuesday. You are sitting in your office—bored. You, a consulting electrical engineer, and nobody has called. Don't they know that you came from the best engineering, sci.... Ring!

Somebody wants me! Ring! I hope they aren't selling vinyl siding. Ring! Better answer it. "Electrical Engineering Consultants!"

It's the accounting department of a small school on the Wabash River about two hours north of your old place. They've been looking at their power bill from Partial Service Electric and Gas and wondering if they can't do something to save some money.

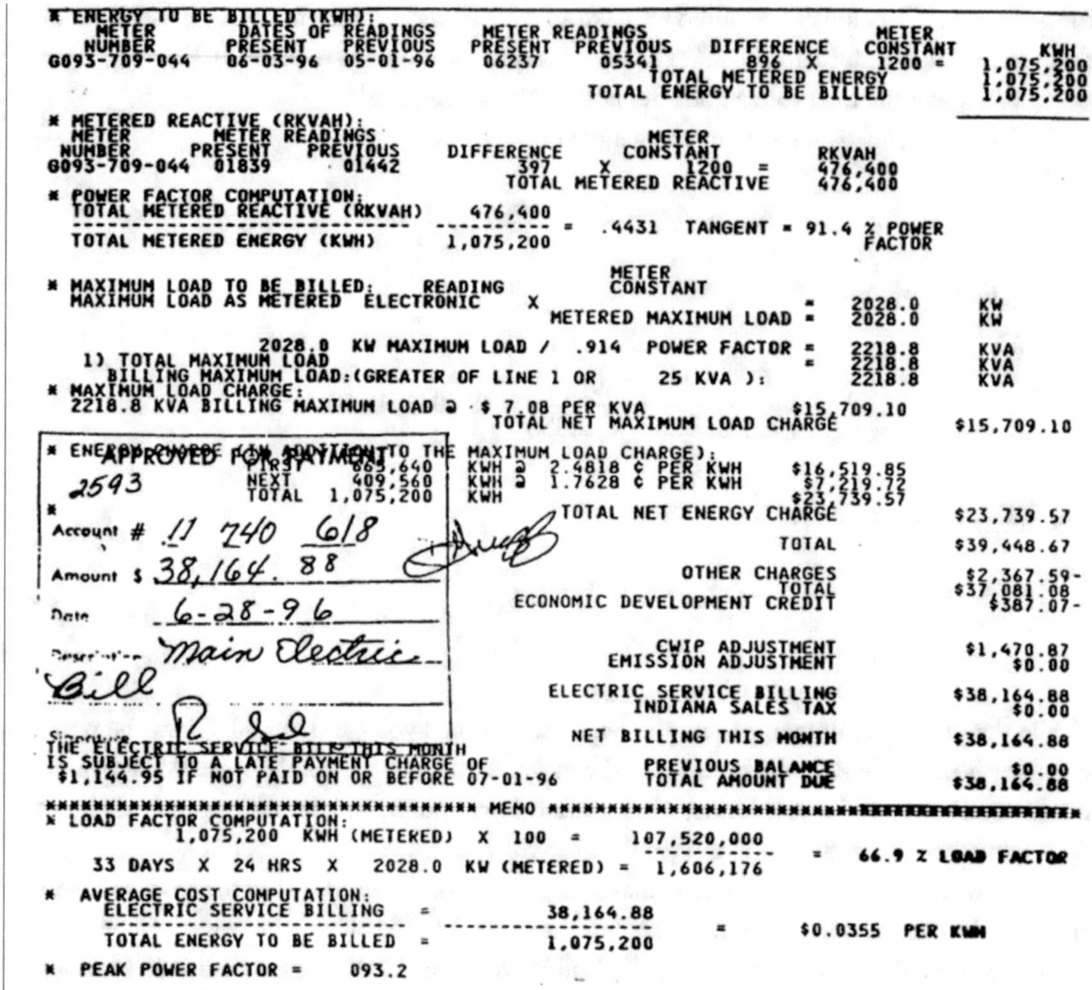

```
* ENERGY TO BE BILLED (KWH):
  METER        DATES OF READINGS    METER READINGS                       METER
  NUMBER      PRESENT   PREVIOUS   PRESENT  PREVIOUS  DIFFERENCE        CONSTANT          KWH
G093-709-044  06-03-96  05-01-96   06237    05341         896  X          1200  =      1,075,200
                                                    TOTAL METERED ENERGY                1,075,200
                                              TOTAL ENERGY TO BE BILLED                 1,075,200

* METERED REACTIVE (RKVAH):
  METER        METER READINGS                            METER
  NUMBER      PRESENT   PREVIOUS    DIFFERENCE          CONSTANT            RKVAH
G093-709-044   01839     01442         397   X           1200  =          476,400
                                      TOTAL METERED REACTIVE              476,400

* POWER FACTOR COMPUTATION:
  TOTAL METERED REACTIVE (RKVAH)      476,400
  -----------------------------  ------------- =  .4431   TANGENT = 91.4 % POWER
  TOTAL METERED ENERGY (KWH)        1,075,200                                 FACTOR

                                                  METER
* MAXIMUM LOAD TO BE BILLED:        READING     CONSTANT
  MAXIMUM LOAD AS METERED  ELECTRONIC    X                        =      2028.0      KW
                                    METERED MAXIMUM LOAD  =              2028.0      KW

            2028.0  KW MAXIMUM LOAD /  .914  POWER FACTOR =        2218.8      KVA
  1) TOTAL MAXIMUM LOAD                                     =      2218.8      KVA
     BILLING MAXIMUM LOAD:(GREATER OF LINE 1 OR    25 KVA ):        2218.8      KVA
* MAXIMUM LOAD CHARGE:
  2218.8 KVA BILLING MAXIMUM LOAD @ $ 7.08 PER KVA         $15,709.10
                              TOTAL NET MAXIMUM LOAD CHARGE             $15,709.10

* ENERGY (IN ADDITION TO THE MAXIMUM LOAD CHARGE):
        FIRST    665,640   KWH @  2.4818 ¢ PER KWH       $16,519.85
        NEXT     409,560   KWH @  1.7628 ¢ PER KWH        $7,219.72
        TOTAL  1,075,200   KWH                           $23,739.57
                              TOTAL NET ENERGY CHARGE                  $23,739.57

                                              TOTAL                    $39,448.67
                                      OTHER CHARGES                    $2,367.59-
                                              TOTAL                   $37,081.08
                              ECONOMIC DEVELOPMENT CREDIT               $387.07-

                                      CWIP ADJUSTMENT                  $1,470.87
                                 EMISSION ADJUSTMENT                       $0.00

                        ELECTRIC SERVICE BILLING                     $38,164.88
                             INDIANA SALES TAX                            $0.00

                        NET BILLING THIS MONTH                       $38,164.88

THE ELECTRIC SERVICE BILL THIS MONTH    PREVIOUS BALANCE                   $0.00
IS SUBJECT TO A LATE PAYMENT CHARGE OF  TOTAL AMOUNT DUE               $38,164.88
$1,144.95 IF NOT PAID ON OR BEFORE 07-01-96

*************************************** MEMO ********************************************
* LOAD FACTOR COMPUTATION:
          1,075,200  KWH (METERED)  X  100  =     107,520,000
          -------------------------------------   -----------   =  66.9 % LOAD FACTOR
     33 DAYS  X  24 HRS  X  2028.0  KW (METERED) =   1,606,176

* AVERAGE COST COMPUTATION:
     ELECTRIC SERVICE BILLING        =        38,164.88
     -----------------------------   ------   ------------   =   $0.0355  PER KWH
     TOTAL ENERGY TO BE BILLED       =       1,075,200

* PEAK POWER FACTOR =    093.2
```

Handwritten overlay on bill:
```
APPROVED FOR PAYMENT
2593
Account # 11 740 618
Amount $ 38,164. 88
Date 6-28-96
Description  Main Electric
             Bill
Signature
```

FIGURE 8.28: PSE&G bill.

Great, you tell them, I will be happy to look at your problem. Dollars dance before your eyes. This place is well funded.

Fax me the bill, you tell them, then…oops, you forget you haven't been able to afford a fax yet. No, this is too important. You'd like to talk with them, so you'll drive over and see what the problem is.

When you get there, you find they have the latest bill from PSE&G. It looks very complicated. (See Fig. 8.28.) But let's see what the problem is.

Below the headings is the line that tells actual energy delivered. This bill appears to be for one month, from May 1 to June 3. When the meter constant is thrown in, the school has used 1,075,200 kWhr of electricity. So far so good.

But PSE&G also meters reactive "energy." That's shown in the second group of lines as 476,400 kVARhr. (The power company hasn't heard of the correct SI units and calls this RKVAH.)

From this they have computed the average power factor:

$$\cos\tan^{-1}\frac{metered\ reactive}{metered\ energy}=\cos\tan^{-1}\frac{476400}{1075200}$$
$$=91.4\%.$$

It's what they do with this that gets confusing. There's another meter that keeps a record of the peak power (kW, not kWhr) usage during the month. The next lines of the bill show that this peak is 2028.0 kW. Using this number and the power factor, PSE&G figures the peak "demand" (*apparent power*) during this period is 2218.8 kVA:

$$peak\ demand=\frac{max.load}{pf}=\frac{2028.0}{0.914}$$
$$=2218.8kVA.$$

Now comes the computation of the amount of money the school owes:
- The demand charge is $7.08 per kVA of demand, with a 25-kVA minimum.
- The first 665,640 kWhr used per month are billed at 2.4818¢ per kWhr.
- The remaining 409,560 kWhr are billed at 1.7628¢/kWhr.
- The total bill is

$$\$7.08\times2218.8=\$15,709.10$$
$$\$0.024818\times665,640=\$16,519.85$$
$$\$0.017628\times409,560=\$\ \ 7,219.72$$
$$Total\ bill=\$23,739.57$$

Can you save them money, they ask? You negotiate a fee, give them a possible time schedule, and head for your office. All sorts of things are going around in your head... energy... power factor... demand... reactive power... VARs... power-factor correction... capacitors.

Aha! That's it! The small school north on the Wabash can make some differences in their bill by practicing energy conservation. You can make a difference by reducing their demand charges.

You look at the bill again. Hmmm, that $7.08 charge per kVA would be minimized by making the power factor larger. After all, it is in the denominator of the computation. But you can't reduce the "demand" number below the peak kilowatts.

OK, this means we need to add enough capacitive load to exactly balance out the metered reactive power. How many kVAR is this? Well, the bill is for 33 days, and there are 24 h in a day, so the reactive load on the average must be

$$\frac{kVARhr}{33 \times 24} = \frac{476,400}{792} = 601.5\,kVAR.$$

This says we need, in round numbers, 600 kVAR of capacitors to balance the school's reactive load. You check with a manufacturer to see what this would cost. They say, $25 per kVAR plus an equal amount for installation. So capacitors cost about $50 per kVAR installed.

Just to be sure of yourself, though, you call one of your old professors at that best engineering, sci… school. After you explain what you've done, she says that it is not a good idea to move the power factor to unity. Since the school's load varies, it's possible at times that the overall load would become capacitive, and a leading power factor can give voltage-regulation problems. She recommends no more than 95 or 96%.

You choose 96%. How much capacitive reactance must you install to get there? You need to know how much reactive power you will have if the power factor is 96% lagging:

$$\cos \tan^{-1} \frac{new\,kVARhr}{1075200} = 0.96,$$
$$new\,kVARhr = 1,075,200 \tan \cos^{-1} 0.96$$
$$= 313,600\,kVARhr,$$
$$new\,kVAR = \frac{313,600}{33 \times 24} = 396.0\,kVAR.$$

The amount of capacitance to add is the new kVAR minus the current value, so

$$capacitor\,kVAR = 396.0 - 601.5 = -205.5\,kVAR.$$

You look in catalogs and the closest capacitor bank you can find at the specified voltage is 198 kVAR. So it looks like the installed cost of this bank is

$$\$50 \times 198 = \$9,900.$$

Is this worth it? First, we need to know what this amount of money, paid "up front," becomes on a *capital-recovery basis*. You recall that an amount P paid out now requires payments according to the capital-recovery formula:

$$A = P\left[\frac{i(1+i)^n}{(1+i)^n - 1}\right]$$

where i = interest rate per period

and n = number of periods.

You check and find that these capacitors can be amortized over ten years, so that's 120 payments, at an interest rate of 7% per annum. So the monthly payments for the capacitors will be

$$A = 9900\left[\frac{0.07/12(1+0.07/12)^{120}}{(1+0.07/12)^{120} - 1}\right]$$
$$= \$114.95 \text{ per month.}$$

How much will this save? You are recommending 198 kVAR of capacitors, which will be a reactive load of

$$-198 \times 33 \times 24 = -156{,}816 \text{ kVARhr.}$$

This reduces the overall reactive load and changes the power factor to

$$\cos\tan^{-1}\frac{476400 - 156816}{1075200} = 95.9\% \text{ lagging.}$$

The maximum demand is reduced because the power factor has increased, so the amount saved will be $708 times the difference in the maximum loads before and after your change:

$$peak\,demand = \frac{max\,.load}{pf} = \frac{2028.0}{0.959}$$
$$= 2114.7\,\text{kVA,}$$
$$saving = \$7.08(2218.8 - 2114.79)$$
$$= \$737.03!$$

You are overjoyed with the obvious result! Your scheme will save that school over $700 per month! And for a cost of only $115 per month!

You are just about to write your report when…Ring! It's that school again. They suggest that you might like to see a whole year's worth of bills to help decide what might be best over a long time. Fax it! you say, because with the monetary rewards you are expecting from this job you've bought a fax machine.

When the figures arrive, though, your heart sinks. The power factor is very seasonal, varying from a low of 91.2% in July a year ago to a high of 97.6% last November.

How good is your correction? What do you do now? And just for the record, you find that the school already has 300 kVAR of capacitors installed in their system to help with the power factor.

What's your next step?

8.6   SUMMARY

(Yes, I know, I left that last example hanging!)

Where have we been in this chapter? After a short review of the phasor domain and the sinusoidal steady state, we've spent our time on power in AC circuits. In other words, we've been working with steady-state power.

What's new in AC power? Four things: real or active or average power (which represents the net transfer of energy), reactive power (which isn't really power at all), complex or total power (the complex sum of real and reactive power), and power factor (the cosine of the angle between real power and complex power or between voltage and current).

Manipulating these quantities isn't difficult if we keep in mind that they are all complex values and must be treated as such. The power triangle is a way of seeing the relationships.

In this chapter we have pretty much restricted our attention to AC power and hence to a frequency of 60 Hz (at least in North America).

Where do you go from here? There are very few places in electrical engineering where you can work and yet avoid dealing with circuits. I hope that what you have learned here will give you a foundation for understanding those circuits. I also hope that you'll be able to create circuits when they are needed.

So I'd like to finish with my original hope that I stated in the first chapter of *Pragmatic Circuits I*, this time with an addition. I will be happy if you can write the mathematical description of a circuit reliably 100% of the time, and that you can carry your knowledge of analysis over into synthesis and design good circuits.

Now we've finished *Pragmatic Circuits II–Frequency Domain*, where we have covered the use of Laplace transforms to transform circuits into the s domain, circuit analysis in the s domain, the s plane, the sinusoidal steady state as a special case where $s = j\omega$, and AC power. What comes next?

Pragmatic Circuits III–Signals and Filters continues in the frequency domain, beginning with signals, their spectra, and Fourier series. It includes the design of active filters and concludes with an introduction to the Fourier transform.

Biography

Bill Eccles has been Professor of Electrical and Computer Engineering at Rose-Hulman Institute of Technology since 1990 (except for one year at Oklahoma State). He retired in 1990 as Distinguished Professor Emeritus after 25 years at the University of South Carolina. He founded the Department of Computer Science at that university, and served at one time or another as head of four different departments, Computer Science, Mathematics and Computer Science, and Electrical and Computer Engineering, all at South Carolina, and Electrical and Computer Engineering at Rose-Hulman. Most of his teaching has been in circuits and in microprocessor systems. He has published Microprocessor Systems: A 16-Bit Approach (Addison-Wersley, 1985) and numerous monographs on circuits, systems, microprocessor programming, and digital logic design. Bill learned circuit theory at M.I.T. under Ernest Guillemin, one of the pioneers in modern circuit theory, and William Hayt at Purdue University. Bill and his wife Trish have two children and three grandchildren. Bill is also a conductor (appropriate for an electrical engineer) on the Whitewater Valley Railroad, a tourist line in Connersville, Indiana. He is a Registered Professional Engineer and an amateur radio operator.

Printed in the United States
by Baker & Taylor Publisher Services